Climate Changes during the Holocene and their Impact on Hydrological Systems

It is now widely accepted that increasing concentrations of greenhouse gases in the atmosphere are affecting the Earth's radiation balance, resulting in higher global atmospheric temperatures. However, there is still a great deal of uncertainty about the likely effects of such a temperature rise on climate, and even more about the impacts of climate change and variability on the world's hydrological regimes and socio-economic systems. Studying the effects of climate variability in the past can give clues regarding possible future effects.

This volume provides a comprehensive review of the effects of climate variability on hydrological and human systems in the Holocene (approximately the last 10,000 years of pre-history and history), in various parts of the world. The book concentrates on the regions bordering the Mediterranean Sea to the east and north, the western and central parts of Europe, China, Japan, west and south Africa and the southwestern USA. The main conclusion is that global warming will bring about a decrease in precipitation in the regions dominated by the westerlies (Mediterranean climates) and an increase in precipitation in the monsoon (sub-tropical and tropical climates) regions.

Climate Changes during the Holocene and their Impact on Hydrological Systems will be of value to researchers and professionals in hydrology, climatology, geology and historical geography.

ARIE S. ISSAR received his M.Sc. and Ph.D. in geology at The Hebrew University of Jerusalem. From 1961 to 1964, he was an advisor to the Government of Iran in the field of groundwater, research and development, under the auspices of the UN Food and Agriculture Organization (FAO). From 1965 to 1974 he was Head of the Hydrogeological Division of the Geological Survey of Israel. In 1975, he founded and later led the Water Resources Research Center of the Jacob Blaustein Institute for Desert Research, and became a Professor of Ben-Gurion University of the Negev, where he was holder of the Alain Poher Chair in the Hydrogeology of Arid Zones. Professor Issar has published over a hundred papers in scientific journals concerning hydrogeology and hydrochemistry of arid zones, the water problems of the Middle East, as well as the impact of climate change on hydrology and socio-economic systems. He has authored or co-authored several books, including:

Groundwater Recharge: A Guide to Understanding and Estimating Natural Recharge (1990; Lerner, D. N., Issar, A. S., Simmers, I. IAH & Verlag Heinz Heise).

Water Shall Flow from the Rock: Hydrogeology and Climate in the Lands of the Bible (1990; Springer-Verlag)

Runoff, Infiltration and Subsurface Flow of Water in Arid and Semi-arid Regions (1996; Edited by A. S. Issar and S. D. Resnick; Kluwer Academic Publishers).

Diachronic Climatic Impacts on Water Resources (1996; edited by A. N. Angelakis and A. S. Issar, NATO ASI Series, Springer-Verlag)

Water, Environment and Society in Times of Climate Change (1998; edited by A. S. Issar and N. Brown; Kluwer Academic Publishers).

INTERNATIONAL HYDROLOGY SERIES

The **International Hydrological Programme** (IHP) was established by the United Nations Educational, Scientific and Cultural Organisation (UNESCO) in 1975 as the successor to the International Hydrological Decade. The long-term goal of the IHP is to advance our understanding of processes occurring in the water cycle and to integrate this knowledge into water resources management. The IHP is the only UN science and educational programme in the field of water resources, and one of its outputs has been a steady stream of technical and information documents aimed at water specialists and decision-makers.

The **International Hydrology Series** has been developed by the IHP in collaboration with Cambridge University Press as a major collection of research monographs, synthesis volumes and graduate texts on the subject of water. Authoritative and international in scope, the various books within the Series all contribute to the aims of the IHP in improving scientific and technical knowledge of fresh water processes, in providing research know-how and in stimulating the responsible management of water resources.

INTERNATIONAL HYDROLOGY SERIES

Climate Changes during the Holocene and their Impact on Hydrological Systems

Arie S. Issar

The Hebrew University of Jerusalem

PUBLISHED BY THE PRESS SYNDICATE OF THE UNIVERSITY OF CAMBRIDGE
The Pitt Building, Trumpington Street, Cambridge, United Kingdom

CAMBRIDGE UNIVERSITY PRESS
The Edinburgh Building, Cambridge CB2 2RU, UK
40 West 20th Street, New York, NY 10011-4211, USA
477 Williamstown Road, Port Melbourne, VIC 3207, Australia
Ruiz de Alarcón 13, 28014 Madrid, Spain
Dock House, The Waterfront, Cape Town 8001, South Africa

http://www.cambridge.org

First published 2003

Printed in the United Kingdom at the University Press, Cambridge

Typeface Times 9.5/13 pt   *System* LATEX 2_ε [TB]

A catalogue record for this book is available from the British Library

ISBN 0 521 81726 9 hardback

To Margalit with love

Contents

Figures and tables

TABLES

Preface

There is a general agreement among scientists that the surface temperatures of both oceans and continents are rising. It is also agreed that greenhouse gases like carbon dioxide and methane are increasing in the atmosphere and that this increase is a result of the continuous rise in human industrial and transportation activity, depending on the fossil fuels, i.e., coal and petroleum. There is still an ongoing debate whether all three phenomena are interconnected and whether part of the blame for the warming should be apportioned to natural processes, such as those that caused climate changes before the industrial revolution. The majority of scientists will not contest natural processes as a possible additional factor but will put the main blame on the emission of greenhouse gases, while admitting that there are some questions which still remain to be solved: such as what is the cooling effect of other products of industry emitted into the atmosphere (e.g., smoke and sulfurous particles, which may cause a shading layer with a cooling effect).

One of the most important tools for investigating the reasons for the global change, as well as for predicting future developments, is computerized general climatological models (GCM), which simulate the physical processes taking place in the atmosphere and beyond, and their impact on the temperatures of sea and land. However, as with all computer models, the correct output is a function of correct input, when input in this case involves data as well as procedures. Simulation–calibration runs are essential for testing procedures. The enormous complexity of the climatic events and their scale requires as many simulation runs as possible, all on the basis of enormous amounts of observations. Meteorological observations are limited in both space and time, especially those based on reliable instruments and observers. This is an obstacle in the use of GCMs to simulate already observed scenarios, not to speak of forecasting future ones. Consequently, when one is considering processes that go back into the past beyond the periods for which data exists or over regions where direct past meteorological data are not available, simulation has to rely on proxy-data. These include measurements that provide clues of past impacts of climate changes on the environment, such as deposits of glaciers, imprints of ancient shores of oceans and lakes and the nature of the deposits in aquatic environments. A major advance in reconstructing ancient climates was achieved by the understanding of the impact of atmospheric and oceanic temperatures on the distribution of environmentally stable isotopes such as oxygen-18, hydrogen-2 (deuterium) and carbon-13. This achievement added an additional dimension to the investigation of ice and sea bottom cores as well as to that of tree rings and cave stalagmites.

While the reliability of the predictions of the GCMs is of general global importance, for certain regions it is crucial; these regions are mainly the low-lying coastal plains bordering the oceans and seas because warming will most probably induce the melting of glaciers and cause the rise of sea levels, flooding these regions. Other regions that will be seriously affected occur along the margins of the desert belts of the globe, where climate change may spell the advance or retreat of the desert, with devastating floods or droughts. The latter problem stimulated me, a hydro-geologist involved in the investigation of water resources in arid zones, to extend my field of investigations into the past. While the basic principle of geology, as determined by the founders of this science during the nineteenth century, is that present processes of erosion and deposition are the key for understanding the past, the maxim of the present study was that past climatological and environmental scenarios are the key for forecasting future events. For this purpose, I searched for paleo-hydrological clues to determine whether processes of flooding and desertification in the past were a function of climate changes or human activity. A good example is the desertion of the cities and the agricultural farms that flourished during Nabatean, Roman and Byzantine times in the arid part of the Levant. Most contemporary archaeologists, historians and ecologists maintain that this desertion was for anthropogenic reasons. When this investigation was progressing and the natural cause became more and more convincing, I was invited by the Division of Water Sciences of UNESCO to join the working group set up by this organization in the framework of the International Hydrologic Program (IHP) to evaluate the impact of climate change on the hydrological cycle. The question was whether data on climate changes in the Middle East, and the impact of such changes in the past on the availability of water resources, could be used to assess the impact of future global change. This invitation was accepted, and indeed study of the data available showed that conclusions from the past could be drawn, as will be discussed later on.

The major climate changes during the Holocene that influenced the history of the eastern Mediterranean region were later correlated to proxy-data time series suggested by different investigators for different regions of the globe. Conformities and non-conformities with the Levant base section were investigated in order to find out whether differences result from variations in the interpretation of data or whether they reflect differences in the nature of the impact of climate change. While trying to do this correlation, one had to take into consideration not only the differences in climate between the various regions but also the vulnerability of the ecological and socio-economical systems in the different areas. Systems along the margins of climate belts were more vulnerable to changes than those in the center of such regions. This difference influenced not only the intensity of the change, if it occurred at all, but also the duration of the impact. For example, regions along the margins of deserts, like the Levant, were first to show the impact of climate change, followed by more humid regions, like Europe, and then by the tropical zones. On top of these differences, one had to take into consideration limitations dictated by the nature of the proxy-data. For example, sea-level changes could be interpreted as either tectonic or eustatic, and changes in the palynological assemblages could be interpreted as either anthropogenic or natural. Yet, notwithstanding these limitations, a rather detailed paleo-climatic columnar section for the Holocene developed, which was able to withstand many tests of prediction. One such test was the forcing of monsoons during warm periods, which was found later to be in agreement with the forecasts derived from the GCMs.

When it comes to the impact of human societies on the environment, it appeared in most cases that it was a severe climate change which decided the history of the environment, rather than human faults. Even so, enough blame for destruction of natural bio-systems and environment still rested on human shoulders. On this basis, I cannot advocate the rejuvenation of the classical geographical "deterministic paradigm", which put all blame on nature and which was endorsed by the geographer Elsworth Huntington and his school during the first few decades of the twentieth century.

Rather, I would suggest a neo-deterministic approach, which considers the human socio-economic and the natural systems as interdependent parts in a general system sensitive to climate changes (Issar and Zohar, 2003). These changes and their impacts can be traced using paleo-environmental proxy-data, such as isotopes, sediments and sea and lake levels. Historical records on changes pertaining to the human socio-economic systems can then be correlated and help to draw more objective conclusions regarding the past. This improves our ability to use the events of the past as a tool for predicting the future of the hydrological systems in periods of global change.

In the following chapters, proxy-data of time series are presented using before present (BP) as the age parameter even though for many of the ages the conventional BCE (or BC) and ACE (or AD) would have been more appropriate for correlation with historical periods and events. However, the historical timetable is in the first place confined to a country or a culture, and the accuracy required is in the order of magnitude of decades if not a few years. For achieving this, the historical timetable is based mainly on archaeological data, which in some cases is based on carbon-14 dates, while in others it is based on pottery stratigraphy as well as historical documentation. By comparison, the dates of the various time series presented in the following chapters are based on several different methods, which differ one from the other in their precision. The levels of precision demanded are that of more than a few centuries during the lower half of the Holocene and a century to a few decades during its upper half. Moreover, as this book is attempting to correlate between events that occurred "simultaneously" in different regions placed in different climate belts, the effect of the climate changes on the environmental time series may be different in its initiation, its duration and its impact. Consequently, the time boundaries of "simultaneous" event climate changes are rather blurred and the level of precision provided by the various dating methods used in the numerous investigations cited in this book were found to be sufficient for inter-regional correlation. The reader should thus regard the term BP, unless the method of dating and precision is defined, as indicating "years ago" and rather flexible according to the range on the time dimension: a few centuries either way during the first half of the Holocene, and a few decades during the second half.

Acknowledgements

The sponsorship and encouragement of the Division of Water Affairs of UNESCO, and especially that of Dr. Michael Bonnel, are gratefully appreciated. The sponsorship of the Jacob Blaustein International Center, which furnished a fellowship to Dr. H. Bruins to summarize part of the material, the help of Mrs Dorit Makover-Levin M.Sc., who collected, organized and interpreted archaeological data from the Levant, and the technical help of Eng. Morel Wolff, who carried out the processing of the graphical data, are warmly acknowledged.

The research reported in the following chapters was supported by colleagues in various universities and scientific institutions, such as as the Institute of Geography of the Chinese Academy of Sciences. Here, discussions with, and material supplied by, Prof. Zhang Peiyuan were most helpful in understanding the impact of climate changes on the region under monsoon regime as well as on its desert continental margins. In this respect, the fellowship grant by the Japanese Society for the Progress of Science, (J.S.P.S), which enabled me to visit the University of Tokyo is thankfully acknowledged. Thanks are also due to Prof. Yanukura and Dr. Tadashi Ogouchi, who supported this project, and to Prof. H. Suzuki, Prof. Y. Sakaguchi and Dr. Aoki for the invaluable discussions that were held with them. Thanks are also due to the Head and to other members of the Department of Geography, Faculty of Science of the University of Tokyo, as well as to its secretary, whose warm hospitality made the stay in the department most enjoyable and fruitful. The technical help of the librarian of the department, Ms. T. Yoshida, is much appreciated.

Thanks are due to the members of the Faculty of Earth Sciences of the Free University of Amsterdam, Dr. Bohncke, Prof. Zagwein, Prof. Dr. Roeleveld, Prof. Dr. J. de Vries, Dr. Isarin, Dr. Krook, Dr. Schwabe and, especially, its Head, Prof. Dr. I. Simmers, as well as to the secretariat of the faculty, especially Miss Alwien Prinsen. My thanks to Prof. S. Jelghersma and Dr. Van der Valk from the National Geological Survey and the NRW for supporting the research carried out in the Netherlands.

The support of the Institute for the Study of the Planet Earth (ISPE) of the University of Arizona, Tucson, and the Department of Hydrology and Water Resources, as well as the help and information supplied by Prof. Nathan Buras, Dr. Lisa Graumlich and Prof. Owen K. Davis, are thankfully and warmly acknowledged.

The hospitality and help of Prof. G. (Mini) Garzón Heydt, Dr. Juan de Dios Centeno and Prof. Ramon Llamas of the University of Complutense, Madrid were inestimable. The information supplied by Prof. Carine Zazo, Madrid, Ramon Pérez-Obiol of the Autonomous University of Barcelona, Dr. José S. Carrión Garcia from the University of Murcia and Dr. Blanca Ruiz Zapata from Universidad de Alcala, Spain is highly appreciated.

COPYRIGHT AND AUTHOR ACKNOWLEDGEMENTS

G. Arnold. Fig. 2.2. Dragoni, W. (1998). Some considerations on climatic changes, water resources and water needs in the Italian region south of 43° N. In *Water, Environment and Society in Times of Climate Change,* eds. A. Issar and N. Brown, pp. 241–272. The Netherlands: Kluwer Academic.

Fig. 2.5. Magny, M. (1992). Holocene lake-level fluctuations in Jura and the northern subalpine ranges, France: regional pattern and climatic implications. *Boreas,* 21, 319–334.

Fig. 4.2. Talma, A. S. and Vogel, J. C. (1992). Late Quaternary paleotemperatures derived from a speleothem from Cango Caves, Cape Province, South Africa. *Quaternary Research,* 37, 203–213.

Fig. 5.2. Stine, S. (1994). Extreme and persistent drought in California and Patagonia during Medieval time. Nature, 369, 546–549.

Fig. 5.2. Hughes, M. K. and Graumlich, L. J. (1996). Multi-millenial dendroclimatic studies from the western United States. In *NATO ASI Series*, vol. 141: *Climatic Variations and Forcing Mechanisms of the Last 2000 Years,* eds. P. D. Jones, R. S. Bradley and J. Jouzel. Berlin Springer Verlag .

1 Climate changes in the Levant during the Late Quaternary Period

At a rather early stage of the research to be reported in this book, it was decided to use the connections between climate changes, hydrological and socio-economic systems in the Levant in order to establish a basic reference sequence of climate changes during the Holocene. Once this had been accomplished, this sequence would be correlated with other regions over the globe. This decision was based on the following observations.

1. This region is a transition zone between two climate belts: the westerlies system and the sub-tropical or intertropical convergence zone (ITCZ) overlying the Arabian–Sahara desert belt. The rate of movement of these two belts north and south affects the mean annual quantity of rain, as well as its variability from year to year. Consequently, the positions in the past of these belts that affect the Mediterranean region's climatic regime and hydrological cycle may provide information reflecting global climate changes.
2. The Nile, which reflects the easterlies and the tropical climate regime over eastern Africa, reaches the Mediterranean and its sediments reflect the history of the climate changes over its watershed.
3. The relatively moderate size of the Mediterranean region, causing climate changes to be rather synchronous (although not absolute) over most of the area, enables establishment of a regional climate change chronology.
4. The long history of human societies in this region, the abundance of documents and archaeological excavations, all facilitate investigation of the impact of climate changes on past socio-economic systems.

1.1 CONTEMPORARY CLIMATE

The Levant is affected by two climate systems. During winter, the westerlies bring in cyclonic low barometric pressures, causing cold air masses to arrive from the Atlantic and the North Sea. These travel over the relatively warm Mediterranean and become saturated by moisture, which is discharged as rain and snow. The rate of movement of the belts southwestward and, therefore, the number, intensity and duration of the rainstorms reaching the region, varies from year to year. When a belt of high pressure remains over the area, rainstorms are less abundant and the year is dry. Because of the configuration of the coastline of the southeastern edge of the Mediterranean Sea, the deserts of northern Egypt, Sinai, the Negev and southern Jordan lie outside the main path of rainstorms approaching from the west.

As can be seen from the multi-annual precipitation map (Fig. 1.1), precipitation usually declines to the south and the east. Yet the topography also has an influence. For example, the rift valleys are in the shadow of the rain coming from the sea and, therefore, are relatively arid, while the mountains receive more rain and snow in winter. The scarcity of rains and the high variance in rainfall from year to year become increasingly great as one goes farther into the desert. Rains in the desert, therefore, are characterized by scarcity and randomness.

Precipitation takes place during the winter months, from November to March. This is an advantage over other regions where rain falls in the summer. The temperatures during the winter are relatively low, which means that evaporation is also low. Consequently, the relative effect of the winter rains is rather high. The development of high-pressure systems often follows the low-pressure systems and causes clear and cold weather conditions. Many of the rainstorms, affected by a barometric low in the northern and central part of this region, enter the desert areas as smaller eddies on the margins of the bigger cone of low barometric pressure. They form small convective cells, a few to tens of kilometers in diameter. This causes rain to fall on a limited area around the center of the cell – other more peripheral areas may remain dry. Such a rainstorm may be of high intensity and last for only a few minutes, or it may continue for up to a few hours. Sometimes precipitation descends as hail, and it may snow at the higher elevations during a cold winter. Rainstorms may be preceded by a barometric high over the desert area. In this case, a flow of dry, hot air from the desert blows dust, which flows in the direction of the barometric low. In the autumn and spring, when dust storms are most abundant, the hot,

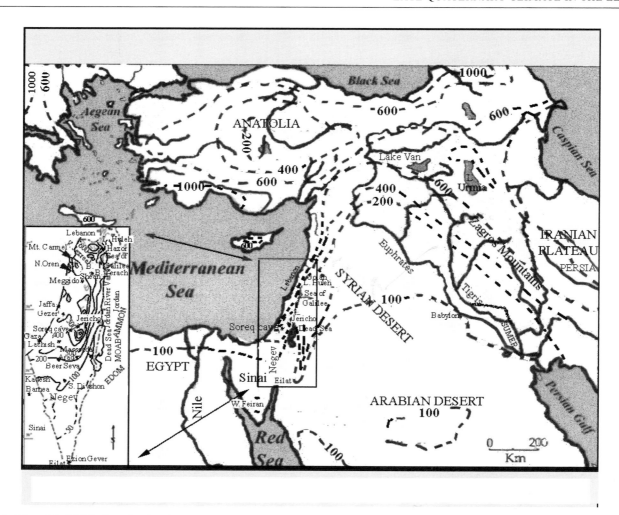

Fig. 1.1. Map of the Middle East showing the multi-annual precipitation (mm per year).

dry periods (known locally as *khamsin*) can come to an abrupt end with a heavy rainstorm. Most dust storms are connected with a barometric high over the continent and lows approaching from the sea.

The Mediterranean Sea acts as a gigantic temperature regulator, because of the high heat capacity of the water. As distance from the sea increases, the regulatory effect decreases. As a result, the temperature differences between day and night, as well as seasonal temperatures, are high. The influence of the Red Sea, the Dead Sea and the Persian Gulf, which are enclosed in narrow depressions, is limited to their very close vicinities. Thus, in the desert areas, the differences between day and night temperatures may reach 15 °C and in some extremes even 20 °C. In summer, the temperature can reach 40 °C during the day, while during the night it drops to about 25 °C. On a winter night, the temperature may fall below 0 °C, while during the day it may reach 20 °C.

Ambient air temperature increases in a directional pattern, similar to that of regional precipitation. (In northern Syria, the average temperature is 5 °C in January and 24 °C in August; in Beirut, it is 13 °C in January and 27 °C in August.)

For inhabitants of these areas, the severity of the high and low temperatures is compensated for by the dryness of the weather during most of the year. This relieves heat stress, since perspiration can evaporate. Humans will feel comparatively comfortable if not exposed to direct sun radiation. However, the dryness causes high evaporation rates from the surface of water bodies and high transpiration rates from vegetation.

During the summer, the weather is less variable, being affected by the semi-permanent surface heat trough centered over Iran and Iraq. This surface trough is coupled with an upper air high-pressure system, producing stable, hot and dry weather. During the autumn (mainly October to November), cool and moist air masses occasionally penetrate the region from the north and produce rainfall. Spring (mainly March to April) is characterized by frequent occurrences of *khamsins* and dust storms, although some rainfall may occur.

Overall, six main air masses, originating over the following areas, affect the weather over the Levant:

1. The Arctic Ocean;
2. The Atlantic Ocean, south and west of Iceland;
3. Northern Russia and Siberia;
4. Northern Russia, being modified while passing over the Volga–Ural basins;
5. The Atlantic Ocean south of the Azores;
6. The North African and Syrian–Arabian desert.

Air masses of the first four areas originate at high latitudes and are characterized by low temperatures and dryness. The masses acquire moisture as they pass over the Mediterranean Sea. The last two air masses originate at low latitudes and are characterized by high temperatures, and dryness, which they maintain.

Rainfall in the Middle East, on the whole, has an inverse correlation with temperature, except in areas under the influence of the summer rainfall regime (Crown, 1972). A synoptic analysis of excessive rainfalls in Israel (Amiran and Gilead, 1954) shows that they are the result of an influx of deep, moist and cold polar air into the eastern Mediterranean along meridian trajectories, which makes contact with the warm surface air in a Cyprus low. With the build-up of the Siberian anticyclone as winter progresses, this situation becomes less probable. There is less chance of a strong jet stream forming over central Europe and the Mediterranean that would feed sufficient air into such a rainfall-causing circulation system. Such excessive rains are, therefore, restricted to the beginning of the season, i.e., November or December.

Aridity in the Levant has three general causes (Otterman, 1974):

- separation of the region from oceanic moisture sources owing to distance or topography (rain shadow);
- the existence of dry stable air masses that resist convective currents;
- the absence of a course of events that cause convergence to create unstable air masses and provide the lifting necessary for precipitation.

Zangvil (1979) investigated the temporal fluctuations of seasonal precipitation in Jerusalem during the period 1946/47 to 1953/54. He employed time spectrum analysis and filtering techniques. A prominent peak appeared in the spectrum at a period of 3.0–3.3 years. (Rainfall oscillations in California also show a peak around 3 years.) The most prominent peak in the spectra occurred at 3.3 years at most of the East African stations. A more than average rainfall in east Africa during the main rain period of January to April is probably associated with a more intense Hadley circulation. This circulation causes strong westerlies in the same longitude, resulting in reduced rainfall in the eastern Mediterranean. Zangvil (1979) suggests that there is, perhaps,

a connection between the El Niño southern oscillation (ENSO) and the rainfall in Jerusalem. The ENSO is a world-wide phenomenon, having a dominant period of 3 to 6 years, which corresponds to Jerusalem rainfall oscillations, the first peak at 3.0–3.3 years and the secondary one at 5 years. A similar observation for the eastern part of the Iberian peninsula was found by Rodó *et al.* (1997).

Analyses of the multi-annual trends of variation of precipitation (Alpert *et al.*, 2002; Ben-Gai *et al.*, 1998) have shown that, while there is a general decrease in the overall quantities of precipitation over the Mediterranean region, there is a trend for an increase in the number of rainstorms of high intensity and for either rainier or drier years within the average rainstorms and years.

1.2 THE CLIMATE DURING THE LATE PLEISTOCENE

In general, the climates during glacial periods of the Quaternary, evidenced in the Mediterranean region by sea regressions, were cold, while interglacial and post-glacial periods, evidenced by transgressions, were warm and dry (Horowitz, 1989). In the coastal plains this resulted in the accumulation of black and brown clayey soils in the marshy areas and red loamy soils on the sandstone outcrops. In the mountain areas on the limestone rocks, terra rosa type soils developed. During the interglacial periods, the deposits of sands and the formation of coastal dunes along the coastal plain indicate a warmer and drier climate, as well as an increased supply of sands. These were brought from the delta of the Nile by the Mediterranean counter-clockwise currents (Emery and Neev, 1960; Issar, 1968, 1979; Rohrlich and Goldsmith, 1984). However, during a short period at the climax of the glacial periods, it seems that the climate became dry (Bar-Matthews *et al.*, 1997), possibly because it fell under the influence of the continental high-pressure zone of eastern Europe. The climate during the Last Glacial Period was not different, namely generally cold and humid, except during its climax. During this glacial period, the water found under the Negev and Sinai deserts in the Nubian sandstone layers was recharged. This is evidenced by its carbon-14 (^{14}C) age (which ranges between 30,000 and 20,000 (30 ka and 20 ka) while the oxygen-18 (^{18}O) to deuterium ratios show an Atlantic, rather than a Mediterranean, pattern (Gat and Issar, 1974). By comparing these ratios with the isotopic composition of contemporary rains and their relation to the trajectories of the rainstorms (Leguy *et al.*, 1983), Issar and Bruins (1983) have suggested that during the Last Glacial Period, a west to southwest trajectory of cyclonic lows was dominant. These came over the Mediterranean to reach the Sinai and the Negev, after entering and crossing the Libyan desert and Egypt. These lows intensified dust storm activity, to be followed by torrential rains. This caused the deposition of a loess layer some tens of meters thick (Issar, 1990). In the southern Sinai, shallow lakes extended all along the drainage basin of Wadi Feiran; the ^{14}C dates of the sediments were 24 ka BP (Issar and Eckstein, 1969). At the end of the glacial periods, at c. 15 ka BP, the deposition of loess became considerably less and, instead, the activity of sand dunes was extended. In the sand layers overlying the loess, epi-Paleolithic type tools were found (Goring-Morris and Goldberg, 1990; Issar and Tsoar, 1987; Issar et al., 1989). Geyh (1994), on the basis of isotopic oxygen and carbon in the paleo-water under the deserts of the eastern Mediterranean, came also to the conclusion that the movement southward of the ITCZ can explain the pronounced climatic variations that characterized the transition from the Late Pleistocene Epoch to the Holocene Epoch. When warmer conditions prevailed, the regions governed by the westerlies became drier while the monsoonal regions became more humid (Geyh, 1994).

A calcareous layer is found in the upper part of the loess section all over the northern Negev (Bruins, 1976; Bruins and Yaalon, 1979). It was deposited c. 13 ka BP, according to radiocarbon dating by Goodfriend and Magaritz (1988). Whether this calcareous horizon is synchronous with the deposition of the loess or was formed later needs further investigation. In my opinion, it is epigenetic and the result of flushing of carbonates and sulfates from overlying layers and their deposition at a certain depth during a period of higher summer rains. This is inferred from the composition of the heavy oxygen and carbon isotopes in the stalagmites of Soreq Cave, which rose abruptly from 13.5 ka to c. 11.5 ka BP (Bar-Matthews et al., 1997). The higher ^{13}C/^{14}C ratio points to the increase of C4 type vegetation, while the higher ^{18}O/^{16}O ratio suggests a warmer climate. These two indicators together would indicate a savanna landscape. In such a landscape, the topsoil becomes enriched in salts during the dry period as a result of evapo-transpiration, while during the rain season these salts are partially leached downwards because of the general decrease in precipitation caused by the warmer climate. The fact that the summer rains coming from the Indian Ocean system were abundant during this period is indicated by the freshwater lake deposits in the erosion cirque of Djebel Maghara in northern Sinai (Goldberg, 1977). Abundant arboreal pollen from this period, which was found in the central Negev, is additional evidence for a savanna habitat in a region that at present holds only a few trees along the riverbeds.

During the Last Glacial Period, the paleo Dead Sea, which at that time extended over most of the Jordan Valley and was known as Lake Lisan (Picard, 1943), clearly had a humid period during the Late Pleistocene, resulting in Lisan-type greenish-gray and laminated clay sediments (Neev and Emery, 1967). Lake Lisan proper was first formed c. 70 ka BP and after a few fluctuations it reached its maximum level of approximately 164 m below MSL at c. 25 ka BP. It stayed at this level for about 2000 years and then the

level started to fall until, at c. 10 ka it was approximately 325 m below mean sea level (MSL) (Bartov et al., 2002), or even 350 m below MSL (Begin et al., 1985).

Stiller and Hutchinson (1980), investigating the stable isotopic composition of carbonates of a 54 m core in Lake Huleh, northern Israel, found [18]O data which suggested that no very drastic climatic changes occurred.

Based on palynological data, Van Zeist (1980) claims that from 24 ka to 14 ka BP it was colder and markedly drier than today and from 14 ka to 10 ka BP, there was an increase in temperatures. Many sites suggest a distinct rise in humidity around 14 ka BP.

Pollen diagrams from Lake Zeribar, Kurdistan, Zagros Mountains (El-Moslimany, 1986) show the absence of trees during the last glacial period and the migration of forest into the region between 10 and 5.5 ka BP. This has been interpreted as indicating aridity during the Pleistocene, with gradually increasing precipitation during its late glacial phase and the Holocene. However, the sensitivity of these species (Quercus aegitops and the associated Pistacia atlantica var. mutica and Pistacia khinjuk) to snow and their tolerance of low overall precipitation indicate that higher snowfall, rather than low precipitation, was the reason they did not thrive during the Pleistocene.

Stevens et al. (2001) investigated a core from the same lake and argue that low [18]O values would suggest a relative increase in winter rains rather than overall changes in effective moisture, and vice versa. Also Griffiths et al. (2001) argue for changes in the seasonality of the rains as an important factor in determining the nature of the sediments at Lake Mirabad, which is situated in the same region.

Based on continuous pollen diagrams from boreholes that penetrated the entire Quaternary sequences of the Hula (Huleh) and Dead Sea lakes, Horowitz (1979, 1989) concludes that the Dead Sea served as a continental base level throughout this period. According to Horowitz, the glacial phases in Israel were manifested by periods of somewhat lower temperatures and higher rainfall, some of it in the summer. The interglacials were hot and dry, with Saharan conditions prevailing. The interstadials had the character of a present-day short, rainy winter and a long, dry, hot summer. It is possible that short dry phases might have occurred in Israel at peaks in the glacial phases, but in general, the periods recorded by low sea levels had a wet climate.

Leroi-Gourhan (1974, 1980, 1981), investigating pollen spectra in the Middle East, found that there were fluctuations of wet and dry phases as well as of temperature during the Lower and Middle Würm. The cold–wet maximum seems to be dated around 45 ka BP, while drought conditions characterized the coldest Würmian phase. This probably explains the scarcity of archaeological evidence of occupation between 23 and 19 ka BP. The Late Glacial Period showed some improvements in climate, dated to 17 ka, 13.5 ka and 12 ka BP. Thereafter, a richer and more diversified flora marked the beginning of the Holocene. Leroi-Gourhan maintains that the increase in pastoral and agricultural population densities since 10 ka BP influenced the soils and vegetation. There is enough evidence to allow us to conclude that it became more humid at about 10 ka BP.

Data from the pollen time series from epi-Paleolithic and Neolithic sites in the Jordan valley, including the regions of Fazael and Mallaha, led Darmon (1988) and Leroi-Gourhan (Leroi-Gourhan and Darmon, 1987) to suggest the following climate changes for the transition period from the Pleistocene to the Holocene:

1. Kebaran (*c*. 19 ka–14.5 ka BP): slightly humid;
2. Geometric Kebaran (*c*. 14.5 ka–12.5 ka BP): a humid period;
3. Natufian (*c*. 12.5 ka–10.3 ka BP): a humid period in the Early Natufian, but the climate progressively becoming drier through the end of the Natufian period;
4. Pre-pottery Neolithic ((PPN) A: *c*. 10 ka to 9.5 ka BP): wetter, marked development of trees, *c*. 10 ka BP; relatively forested conditions between 10.25 ka and 7.9 ka BP.

Weinstein (1976) investigated the late Quaternary vegetation of the northern Golan, manifested by the pollen assemblage of samples from borehole P/8, drilled at the center of the lake of Birket Ram. This is a rather small, elliptical volcanic crater lake, 900 m × 600 m, bordered by very steep slopes. The present average annual precipitation is 1000 mm. The fluctuations in pollen samples seen in this section are significant, and a more intensive dating effort should be carried out since dates are rather scarce. A gradual change from a more forested landscape to a Mediterranean one can be seen in the upper part of the section, from 39 to 30 m (at 36 m, ^{14}C age is 28,400 ± 3000 BP). The assemblage is 80% arboreal pollen, of which conifers constitute 88%. From 30 to 22.5 m, the arboreal assemblage is reduced to 33%, consisting of 75–80% *Quercus* sp. and 40% Irano-Turanian types. From 22.5 to 11.5 m, there is an increase in the arboreal assemblage to 60%, of which 89% is *Quercus* sp. and only 20% Irano-Turanian types. One can conclude that towards the upper part of the profile, presumably uppermost Pleistocene, the climate became more humid.

A geomorphological study was carried out by Sakaguchi (1987) in the district of Palmyra in the eastern arid part of Syria (present mean annual precipitation is 125 mm). This survey provided the evidence for the existence of a pluvial lake, which went through periods of high and low levels since at least 100 ka BP. A wet period of the lake ended *c*. 19–18 ka BP; later it became brackish to saline and totally dried up, leaving behind a sabkha. At 10 ka BP, it rejuvenated and existed until 8 ka BP.

A study by Klein *et al.* (1990), of fossil and modern *Porites* corals from reef terraces in the southeastern Sinai along the Red Sea, indicates that the sea level was higher and a wetter climate

prevailed in Sinai during the Late Quaternary, possibly with a summer rainfall regime. Most fossil corals showed degrees of fluorescent banding after irradiation with long-wave ultraviolet light, while living *Porites* corals did not exhibit distinct fluorescent banding. The source of fluorescence is humic acid of terrestrial origin, as was found in corals from the Great Barrier Reef of Australia (Isdale and Kotwicki, 1987). The distinct fluorescent banding in the fossil Sinai corals is understood to be a function of periodic terrestrial runoff floods during the lifetime of the corals, irrespective of later events. Modern corals show skeletal banding patterns: low-density bands being deposited in summer and narrow high-density bands in winter. Fossil corals have a similar density-banding pattern. An important finding is that the fluorescent bands related to humic acid from runoff floods are superimposed on the low-density portions of the skeleton bands, which implies summer rainfall (Klein *et al.*, 1990). This is in accord with the conclusion, already mentioned, that during warmer periods the climate of Sinai was influenced by summer rains.

According to Herman (1989), surface water temperatures of the Mediterranean, during glacial temperature minima, were *c*. 3 °C lower than the present in summer and *c*. 3–4 °C lower in winter. Salinities were highest during the peak of the glacial period when climates were more arid than today. The sea level was very low (130–140 m below MSL); the discharge of the Nile was greatly reduced and the connection between the Mediterranean Sea and the Black Sea (Bosphorus sill at 36 m below BSL), which is a major supplier of low-salinity water, was reversed.

Thunell and Williams (1983, 1989) investigated the paleo-temperature and paleo-salinity history of the Eastern Mediterranean during the Late Quaternary. They maintain that the Mediterranean isotopic signal is a complex record of regional temperature and salinity changes superimposed on compositional changes caused by the global ice volume effect. Hydrographic conditions in the Mediterranean at 8 ka BP must have been considerably different from those at 18 ka BP as well as from those of today. The water balance at 8 ka BP became positive as precipitation and runoff exceeded evaporation. Salinities were considerably lower at 8 ka BP and the west–east (increasing) salinity gradient was reversed to an east–west gradient. This is supported by east African climate records, which indicate the onset of very humid conditions at *c*. 12.5 ka BP, with wettest conditions occurring between 10 ka and 8 ka BP. This was also a time of intensified African monsoons and increased Nile discharge.

Larsen and Evans (1978) reported on findings from layers of the Hammar Formation in the subsurface of the present delta of the Shat-el-Arab. These contained recent marine fauna. They consider these findings as evidence for a transgression phase starting *c*. 10 ka BP. The fresh and brackish water deposits with marine lenses overlying the Hammar Formation are interpreted as layers laid down in a deltaic environment, caused by the progradation of the delta to the southeast. This has advanced c. 180 km during the last 5000 years

Sanlaville (1992) carried out geomorphological investigations of the paleo-climate of the Arabian Peninsula and found that these four humid phases occurred during the Quaternary. The two earliest stages, between c. 128 ka and 105 ka BP (isotopic stage 5e) and between 85 ka and 70 ka BP (isotopic stage 5a), as well as the last one, which took place during the earlier part of the Holocene, could be correlated with northward movement of the monsoon rains. He attributed the wet inbetween phase, which occurred during isotope stage 2, to a southward migration of the westerlies belt.

It can be concluded that the transition period from the Pleistocene to the Holocene was one of general warming up, but with considerable fluctuations. In general, the frequency and intensity of the typical heavy dust and rainstorms, causing the deposition of the loess, decreased and, instead, the supply of sand and mobility of the sand dunes of the Sinai and Negev increased. This increase in the supply of sand resulted from the higher levels of the Nile and the strengthening of the rainstorm system over eastern Africa, The sand supply to the eastern Mediterranean was, probably, reinforced by the erosion of the Nile delta caused by the rise in the sea level. A warm period characterized by summer rains may be distinguished between 13 ka and 11 ka BP. This may have been followed by a cold humid spell, which more accurate dating may correlate with the Younger Dryas. This was followed by a warmer period, which continued until about 10.5 ka BP.

1.3 CLIMATE CHANGES DURING THE HOLOCENE IN THE LEVANT

The initial procedure adopted by the author to establish the sequence of climate changes during the Holocene was based on a chrono-stratigraphical cross section derived mainly from the sequence of ratios of $^{18}O/^{16}O$, with the ratios of $^{13}C/^{12}C$ as auxiliary data (Fig. 1.2). These isotopic data came from a core from the bottom of Lake Van in Turkey (Lemcke and Sturm, 1997), from a core at the bottom of the Sea of Galilee (Stiller et al., 1983–84), from speleothemes of caves in upper Galilee (Issar (1990) based on M. A. Geyh et al., unpublished data) and from the Soreq Cave in the Judean hills in the central part of Israel (Bar-Matthews et al., 1998a,b) and from cores at the bottom of the eastern most part of the Mediterranean Sea (Luz, 1979; Schilman et al., 2002). Needless to say, each time series has its advantages as well as constraints, especially when it comes to the dating of the various layers. Consequently, the time boundaries suggested in this cross section (Fig. 1.2) should be taken as a synthesis and a marker zone, which may fluctuate on the time dimension either because

of the natural environment or because of the different methods of sampling and dating.

The reason for choosing sequences of ratios of $\delta^{18}O/^{16}O$ (the relative proportion of ^{18}O to ^{16}O in the sampled water compared with the isotopic composition of standard mean ocean water (SMOW)) as the most significant time series was because these ratios are strongly influenced by the ambient temperatures and climate regimes in general (Ferronsky and Polyakov, 1982; Fritz and Fontes, 1980; Gat, 1981) but are not influenced by anthropogenic activities. It was also assumed that, in the Middle East, the influence of climate changes on the $\delta^{18}O/^{16}O$ ratio could have been rather pronounced, based on the observation that the isotopic composition of contemporary rainwater is influenced by the trajectories of the rainstorms (Leguy et al., 1983). There is no reason to suggest that such changes in the global climate regime would not have equally influenced these trajectories, and thus the $\delta^{18}O/^{16}O$, in the past. Therefore, interpretation of the stable isotope data as climate and humidity indicators follows the basic assumption that the $\delta^{18}O$ values of precipitation are interrelated with temperature (Geyh and Franke, 1970) and with other meteorological factors (such as changes in the storm trajectories, in the seasonal distribution of precipitation and humidity (Gat, 1981; Leguy et al., 1983) and higher or lower rates of evaporation). This assumption was indeed justified by the interrelations that could be shown between the isotope time series and other proxy-data time series, as will be shown below.

As already mentioned, when correlation lines are drawn, small discrepancies caused by the dating and time scales used in the different data sources must be taken into consideration. These apply to the different amplitudes of the $\delta^{18}O$ records of the lake and sea sediments and of the speleothemes. For example, the water balance of the Sea of Galilee is also determined by an inflow of groundwater from the flanks of the rift valley. Spring water collected along the shore yielded ^{14}C dates of more than 10 ka BP. This would "dampen" the corresponding isotope variations. A certain retardation factor should be taken into consideration for the isotopic composition of the sediments of Lake Van, where part of its inflow comes from springs. In contrast, changes in $\delta^{18}O$ values of speleothemes reflect the fluctuations of isotope composition of the meteoric water over decades. The samples of 1 mm thickness analyzed represent age ranges of about 10 years.

In addition to the problems involved in the ^{14}C dates in relation to the isochrones, some other elements must be taken into consideration. First, the curves presented in the cross sections are modified by the running average method, in order to reduce the impact of noise created by short-term but intense fluctuations. Second, there are differences caused by the reservoir effect of the non-saturated and saturated zones in the subsurface of speleothemes, which is similar to the effect of groundwater storage for springs. Yet even with all these uncertainties, an apparent general

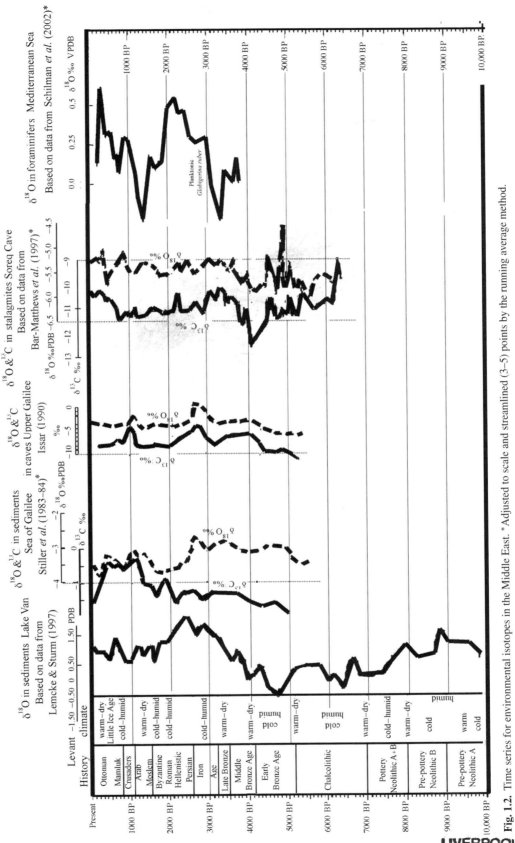

Fig. 1.2. Time series for environmental isotopes in the Middle East. *Adjusted to scale and streamlined (3–5) points by the running average method.

correlation of covariations can be observed. However, because of the problems outlined above, it is suggested that conclusions should also take into consideration other time series of natural proxy-data that are available for this region. These include the paleo-levels of the Mediterranean Sea, the ratios of planktonic foraminifers in the sediments of the eastern Mediterranean, and the Dead Sea lake levels (Fig. 1.3).

As presented in Fig. 1.2, the cross section starts with the $\delta^{18}O$ and $\delta^{13}C$ time series obtained from lacustrine carbonate cores drilled in Lake Van in Eastern Turkey (Lemcke and Sturm, 1997; Schoell, 1978), which is a closed lake at an altitude of 1720 m above MSL. The precipitation on the drainage basin of the lake is influenced by the Mediterranean climate system. The isotopic investigation is part of a general study that has been carried out by a multidisciplinary group (Degens *et al.*, 1984.) The lake has a volume of 607 km^3 and a maximum depth of 451 m and is in a tectonically active zone in eastern Anatolia. The lake level was at its highest at the height of the Last Ice Age, about 18 ka BP when it was 72 m above the present level. According to the pollen analysis, the vegetation was of a steppe type from 10 ka to 6.5 ka BP; from 6.5 ka to 3.4 ka BP, it was forest vegetation and from 3.4 ka BP to the top of the section, the vegetation is contemporary and shows the impact of agriculture. The drop in the level of the lake and the increase in the salinity of the water between 10 and 9 ka BP were interpreted as a change to a warmer and dryer climate. This can be observed in a trend towards a heavier composition of the $^{18}O/^{16}O$ ratios in the isotope curve. Around 7 ka BP, there was a rise in the level of the lake, a decrease in its salinity and a marked increase in the percentage of arboreal pollen. This is interpreted as a change to a more humid climate. One can observe a simultaneous decrease in the $^{18}O/^{16}O$ ratios. At *c.* 3.5 ka BP there is again a sharp decrease in the $^{18}O/^{16}O$ ratio, which reaches its lowest level at 2.7 ka BP and marks another cold period. Because of the increase in agricultural activities since then, the pollen and sedimentological records may present the impact of anthropogenic processes, and the author prefers to rely mainly on the isotope curve, which shows relatively low ratios from 1.8 ka to *c.* 0.8 ka BP, a heavier composition between 0.8 ka and 0.5 ka BP and an increase in the ratio at the top of the column.

Another isotopic composition time series, presented in Fig. 1.2, is that from a core taken from the Sea of Galilee (Stiller *et al.,* 1983–84). This lake is fed by the Jordan River, and by the floods and springs from Galilee and the Golan Heights. Thermal springs also flow into the lake. The base flow of the Jordan is maintained by the outflow of springs emerging from the aquiferous Jurassic limestone rocks of Mount Hermon, in the eastern part of the Anti-Lebanon. These rocks are highly permeable and the water from the rain falling on the mountain and from the melting snow, which covers the higher stretches of the mountain each winter, quickly infiltrates the subsurface to enrich the aquifer from which these

springs arise. The average annual precipitation on the mountains may reach 1200 mm. The two main springs feeding the upper Jordan are the Dan and the Banias (comes from Pan, the Greek god patron of springs). Because of high permeability and the high rate of precipitation, the water flow of these two major springs is fairly regular. The difference between summer and winter is regulated by the large underground storage of Mount Hermon. A long spell of dry years and low snowfall on the drainage basin may cause a decrease in the total quantity of water in the springs, leading to a reduction in the flow of the Jordan and a low water level in the Sea of Galilee. This is intensified by a decrease in the volume of the floods and by higher evaporation rates from the lake, causing the levels of the lake to drop. One may assume that the ratio of $^{18}O/^{16}O$ in the carbonate sediments will be higher in such years. While the precipitation on the catchment area of the springs emerging from the southern tip of Mount Hermon is high, the precipitation on eastern Galilee and the Golan Heights, which form the catchment area of the floods and springs, is less abundant, and the rates of flow are strongly influenced by the average annual rainfall.

The reinterpretation, carried out by the present author, of the $\delta^{18}O/^{16}O$ sequence from this core was correlated with the $\delta^{13}C/^{12}C$, data, assuming that depleted ratios signify more humid conditions, and thus abundant C3 types of vegetation, while a heavier composition indicates a drier climate and abundance of C4 type of vegetation.

Only four ^{14}C dates (at 5240 ± 520, 2955 ± 220, 2170 ± 125, 1020 ± 115 BP) were taken, the oldest one of which was near the bottom of the core hole at *c.* 5.0 m. Nevertheless, the spread of the dated samples along the column, and the body of other proxy-data available, in addition to $\delta^{18}O/^{16}O$, $^{13}C/^{12}C$ ratios (i.e., percentage of CaCO$_3$) and the detailed pollen analysis (Baruch, 1986; Stiller *et al.,* 1983–84), enable this time series to be used to interpret climate changes in the region during the upper half of the Holocene.

These data have been used by the author in his argument against the prevailing paradigm which claims that no significant climate changes occurred during the upper part of the Holocene and attributes all environmental changes to human activity (Issar, 1990). This is also the case with the data from the Sea of Galilee (Stiller *et al.,* 1983–84), which were initially interpreted as reflections of anthropogenic factors rather than climate changes.

The examination of this core (Fig. 1.2) enables us to distinguish various zones. Zones of high $\delta^{18}O/^{16}O$ and $\delta^{13}C/^{12}C$ ratios are from *c.* 5.0 ka to 4.5 ka, from 2.8 ka to 2.3 ka, from *c.* 1.5 ka to *c.* 1.2 ka and, finally, at 0.4 ka BP. Zones with only high $\delta^{13}C/^{12}C$ ratios are from 4.2 ka to 3.5 ka and at 1.8 ka BP. Toward the uppermost part of the $\delta^{18}O/^{16}O$ curve, starting at *c.* 0.3 ka BP, there is a trend to heavier ratios.

The other $\delta^{18}O$ and $\delta^{13}C$ time series presented in Fig. 1.2 average the results of 41 stalagmites taken in 10 caves in Galilee,

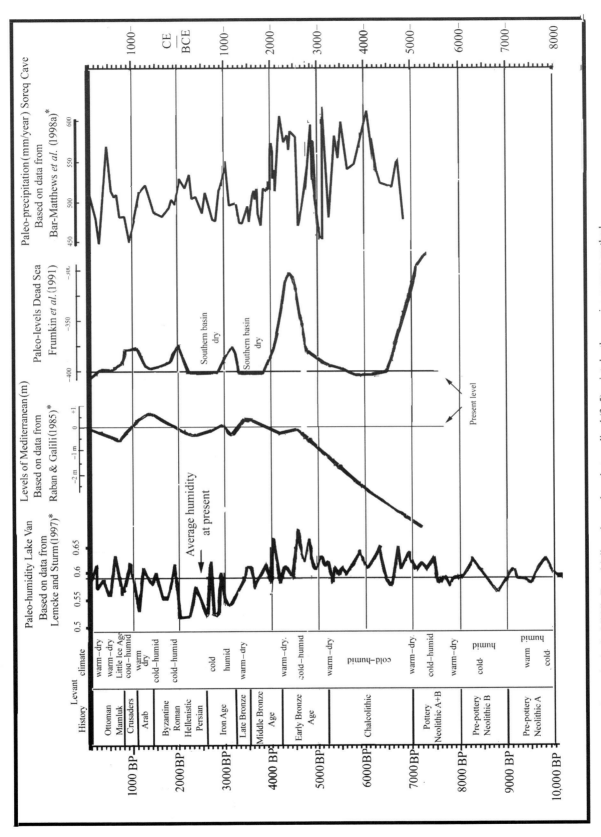

Fig. 1.3. Paleo-hydrology time series in the Middle East. * Adjusted to scale and streamlined (3–5) points by the running average method.

northern Israel. The age determination for all the sequence was from calibration of ^{14}C and uranium/thorium (^{234}U/^{230}Th) dates. Precisions of *c.* 300 years have been obtained, taking a reservoir effect of 900 years into account (Geyh *et al.,* unpublished data). Therefore, speleotheme ages are considered to be calibrated dates with a ± 300 years margin of error. That these dates, within this margin of error, are reliable can be deduced from the close similarity between this speleotheme curve and the one from the Sea of Galilee. Periods of heavy isotopic composition occurred *c.* 4 ka, 3.8 ka and 1.2 ka BP, while periods of light compositions occurred *c.* 4.8 ka, 3.3 ka, 2.0 ka and 1.0 ka BP.

Another sequence of δ^{18}O/^{16}O, forming a time series of paleoclimatic significance, is of a speleotheme from a cave in the vicinity of Jerusalem, in the mountainous part of central Israel (Fig. 1.2; Ayalon *et al.,* 1998; Bar-Matthews *et al.,* 1991, 1993, 1996, 1997, 1998a,b). The age determinations were made by the ^{230}Th/^{234}U method (Kaufman *et al.,* 1998). The isotopic record, which is traceable for the last 58 ka, shows a pronounced difference between the values characterizing the speleothemes that were formed before 6.5 ka BP and those formed later, including the contemporary deposits. This, according to Bar-Matthews *et al.* (1998a,b), is probably because of altogether different climatic regimes.

This is an important observation with regard to the exact time dimension that is suitable to provide proxy-data for simulations using general circulation model (GCM) scenarios. Climate scenarios of the Pleistocene (glacials and interglacials) are not suitable whereas that starting *c.* 6 ka BP is. With regard to the climate changes during the last 6.5 ka, the team working on the speleothemes of Soreq Cave (Ayalon *et al.,* 1998; Bar Matthews *et al.,* 1998a,b) have calculated the paleo-rainfall values by correlating the paleo-δ^{18}O records with the contemporary ratios of δ^{18}O/rainfall. Based on the δ^{18}O and δ^{13}C values and calculated paleo-rainfall, they divide the record into four stages. Stage 1 lasted from 6.5 ka to 5.4 ka BP and was very wet. During the period extending from 5.6 ka to *c.* 3.0 ka BP (stage 2), the climate was, in general, humid, interrupted by four short dry spells. One was between 5.2 ka and 5.0 ka and another was at *c.* 4.0 ka BP. Stage 3, lasting from *c.* 3.0 ka to *c.* 1.0 ka BP, was transitional to drier and more stable conditions. Stage 4, from *c.* 1.0 ka BP to the present, was characterized by fluctuations in rainfall. The high values between 0.4 ka and 0.5 ka BP may be connected with the Little Ice Age, while the increase in δ^{13}C values, which started *c.* 0.7 ka BP, may indicate a process of deforestation and increased grazing during the Turkish period.

The curve of δ^{18}O composition of pelagic and planktonic foraminifers (Luz, 1991) is not too conclusive. In general, changes in the δ^{18}O/^{16}O values reflect changes in oceanic temperatures and the water in which the animals lived. Such ratios in foraminifers' shells in deep-sea sediments enabled the establishment of the sequence of climate changes during the Quaternary (Emiliani, 1955).

This, however, is mainly seen in planktonic assemblages. The isotopic record for the benthonic forms shows low fluctuation because of the relatively stable temperature of the water at the bottom of the sea. Shackleton and Opdyke (1973) have demonstrated that the changes in isotopic values reflect the changes in the continental volume of ice as melted glacier water causes the water of the oceans to become isotopically lighter. It is certain that the fluctuations in the isotopic composition between glacial and interglacial periods resulted from the glacial effects (Bowen, 1991). However, it seems that the isotopic composition of the Mediterranean Sea is more complicated as the isotopic record from cores taken from the Mediterranean Sea also seems to reflect local changes. Consequently the isotopic composition of the Mediterranean Sea reflects not only climatic parameters such as precipitation, evaporation and residence time of water mass within the basin but also the hydrological regimes of the Black Sea, the Nile and the Atlantic Ocean.

From the ratio of δ^{18}O/^{16}O of the epi-pelagic foraminifer *Globigerinoides rubes*, Rossignol-Strick *et al.* (1982) found that the oxygen isotopic composition ratio decreased through several large shifts to minimal values between 8 and 6 ka BP (see also Luz and Perelis-Grossowicz, 1980). The sharp depletion in ^{18}O/^{16}O ratio and the lowering in salinity from 8 ka to 7 ka BP may represent the heavy Nile floods during a mainly rainy period in Africa (Nicholson, 1980; Nicholson and Flohn, 1980), and it seems likely that the Nile was the major source of fresh water responsible for the low salinities in the Mediterranean Sea. However, the Nile water, coming from areas of low latitude, should be isotopically heavy. In a recent work, Luz (1991) suggested an alternative explanation for the isotopic depletion. He claimed that a high influx of low-salinity water entered the Mediterranean Sea from the Black Sea when the rising sea surface reached the level of the Bosphorus. This alternative explanation is not in agreement with the findings of Erinc (1978), who concluded that, even though the sea level rise in the Black Sea and in the Mediterranean Sea started simultaneously after the peak of the last glacial period, the Black Sea basin was disconnected from the Mediterranean. Moreover, the rise in sea levels caused the intrusion of the Mediterranean into the Black Sea. Cores obtained in the Black Sea (Degens, 1971; cited in Erinc, 1978) indicate, "that the main intrusion of saline water into the Black Sea started 7140 ± 180 years ago". In conclusion, it is clear that the reasons for the changes in oxygen isotope composition of the foraminifers of the Mediterranean have yet to be elucidated. In their book *Noah's Flood: The New Scientific Discoveries about the Event that Changed History*, the marine geologists William Ryan and Walter Pitman (1998) argue that this intrusion (around 7500 years ago) was caused by the breaching of the barrier at the Bosphorus, at the northeastern part of the Sea of Marmara, which filled up an ancient lake, the predecessor of the Black Sea, the level of which was 150 m lower than the present sea level. This caused a tremendous waterfall of seawater flooding the lowlands surrounding the ancient lake, a calamity for the people in the Neolithic agricultural communities that lived in this region. They further claim that this calamity lived on in the memory of the people who survived and migrated into Mesopotamia. The stories told were passed on from one generation to the next until they crystallized in the mythological texts found on ancient clay tablets of ancient Mesopotamia. At a later period, these stories were incorporated into the Hebrews' sacred scriptures and became part of the Judeo–Christian–Moslem heritage. Issar and Zohar (2003) maintain that the findings of an ancient flood filling up the Black Sea to its brim is, undoubtedly, of the greatest importance for the understanding of the prehistory of Europe and Central Asia during the Lower Holocene (i.e., 10 ka to 5 ka BP). And yet, this discovery should not be mixed up and confused with the Biblical Flood.

Schilman *et al.* (2001, 2002) examined two cores drilled at the sea bottom in the southeastern part of the Mediterranean near the shores of Israel for oxygen and carbon isotope composition as well as for physical and geochemical properties of the sediments. The date of the lowest layer of the sequence is *c.* 3.6 ka BP. According to Schilman *et al.* (2001, 2002), the $\delta^{18}O$ values of the planktonic foraminifer *G. ruber* suggest that humid phases took place between 3.5 ka and 3.0 ka and between 1.7 ka and 1.0 ka BP, while arid conditions prevailed between 3.0 ka and 1.7 ka BP. At *c.* 0.8 ka BP, a warm period, the Medieval Warm Period, took place and at 0.27 ka BP, a cold period, the Little Ice Age, occurred. Schilman *et al.* (2001, 2002) also suggest a long-term trend of aridization that started *c.* 7.0 ka BP in the mid–low-latitude desert belt and has continued until the present. They base their suggestion on the long-term slight increase in $\delta^{18}O$ values of planktonic foraminifers, which corresponds with a gradual decrease in the $\delta^{13}C$ values of both *G. ruber* and the benthos foraminifers *Uvigerina mediterranea*. This trend is concurrent with an increase in sedimentation rates, the titanium/aluminum (Ti/Al) ratio, magnetic susceptibility and color index of the sediments. Schilman *et al.*, suggest that this general long-term warming up, and thus aridization, reflects a gradual change in the $\delta^{13}C$ of the dissolved CO_2 of the entire southeastern Mediterranean water column, which parallels the global rise of atmospheric CO_2 observed for the late Holocene. They suggest that this is a result of terrestrial biomass destruction during the aridization process and the gradual reduction of the vegetation cover in east Africa, which led "to an increased erratic flood-related sediment flux via the Nile River. This is reflected by the general change in the local sediment composition. At 3.6 ka ago, the Saharan eolian input reached 65% whereas at about 0.3 ka ago 70% of the SE Mediterranean sediment was composed of Nile particulate-matter." I prefer to put more emphasis on the relative fluctuations of the isotopic composition and sedimentary sequence rather than on the general trend.

As can be seen from Fig. 1.2, the two peaks of light oxygen isotopes (*c.* 3.40 ka and 1.4 ka BP), which suggest an influx of melted glacial water (i.e., a warm climate), correspond to two major warm periods: the Late Bronze and the Arab period. A secondary light oxygen period occurred at *c.* 0.7 ka BP, which corresponds to the Mamluk–Ottoman warm phase. The two peaks of heavy oxygen isotopes (*c.* 2.30 ka and 0.3 ka BP) correspond to two cold periods: the Roman and the Little Ice Age. A secondary cold period occurred at *c.* 1.0 ka BP, which corresponded to the Crusader period. I would also interpret the change in the sediment characteristics differently. The higher loess load at *c.* 3.60 ka BP could be the result of the inflow of loess from the higher rate of dust storms and floods in northern Egypt, Sinai and Negev during the Middle Bronze Age (MB), which was relatively (to the Intermediary and Late Bronze periods) cold and humid. The higher sand supply at *c.* 0.3 ka BP would be a function of the general warming up that started at *c.* 1.4 ka BP, which brought higher rates of easterly rainstorms over northeastern Africa and higher supplies of sand from the Nile. This corresponds with the post-Byzantine invasion of sand dunes into the coastal plain of Israel.

The curve reconstructing the sea-level changes along the coastline of central Israel, presented in Fig. 1.3, is by Raban and Galili (1985). It is based on a survey of archaeological sites along the Israeli coastline, with submarine as well as surface structures. It incorporates the results of the work of Galili *et al.* (1988), who reconstructed ancient sea levels between 8 ka and 1.5 ka BP along the coast line of Mount Carmel, and the conclusions of the survey of Bloch (1976), who based his observations on the altitude of ancient salt production basins. The curve of Raban and Galili (1985) shows that, since the lower Holocene, the sea level has risen to reach that of the present day. Between 8 ka and 6 ka BP, the sea level rose at a mean annual rate of 5.2 mm. According to these authors, no tectonic movements have occurred in the area during the last 8000 years. The most pronounced recessions of sea levels shown on this curve are between 4.5 ka and 4.0 ka BP, between 3.5 ka and 3 ka BP, between 2.5 ka and 2 ka BP and *c.* 0.7 ka AD. The periods of high sea level are around 5 ka BP, between 4 ka and 3.5 ka BP, from 3 ka to 2.5 ka BP and *c.* 1.4 ka BP. A trend toward a higher sea level can be seen after 0.5 ka BP. It is suggested that the periods of low sea levels correlate with periods of cold climate, that is, periods of expansion of polar glaciers, while during periods of high sea level the climate was warm and the glaciers melted.

While high levels of the Mediterranean indicate periods of warm climate and vice versa, high levels of the Dead Sea during the Holocene indicate cold and humid periods. This is because the Dead Sea is located at the lower end of the Jordan catchment basin, and its levels are determined by the amount of precipitation on this basin and the rate of evaporation from its surface. The curves of the levels of the lake presented in Fig. 1.3 are based on a survey of ancient shorelines and erosion channels inside the salt caves of Mount Sodom (Frumkin *et al.*, 1991). The results are in agreement with a prior survey that was based on ancient shorelines (Klein, 1982). The periods of high lake levels were found to have occurred *c.* 8 ka, 4.5 ka, 3 ka, 2 ka and 1 ka BP, while periods of very low levels, which most probably caused the drying up of the southern part of the Dead Sea, occurred *c.* 6.5 ka and 2.5 ka BP. In assessing the evidence from these curves, we have to take into consideration the fact that higher and younger lake levels may obliterate the evidence of older but lower levels.

A rather similar pattern of climate changes could be deduced from the carbon and oxygen isotope values investigated in a speleotheme in Nahal Qanah Cave, central Israel (Frumkin *et al.*, 1999).

1.4 CORRELATION BETWEEN CLIMATE CHANGES AND HISTORICAL EVENTS IN THE LEVANT

The climate changes derived from the curves presented in Figs. 1.2 and 1.3 were correlated with archaeological-cultural chronostratigraphy, especially as it relates to the history of the settlements in the desert regions (Issar *et al.*, 1989, 1992; Issar, 1990, 1995a,b; Issar and Makover-Levin, 1995). As will be shown presently, a rather good correlation was found. Consequently, it is suggested that the archaeological chrono-stratigraphy should be used as the paleo-climate stratigraphy of the Holocene for the Eastern Mediterranean region. One must be aware that the dating of the environmental time scale is mainly by ^{14}C methods whereas archaeological dates are also based on the history of civilizations, particularly those that left written documents. In the present work, dating is based also on data presented as non-calibrated ^{14}C ages. The reason for this is that the order of magnitude of correction in the first half of the Holocene is a few hundred years, which is the order of magnitude of the accuracy of the time series of the proxy-data presented in these sections. This lack of accuracy is mainly because the dating intervals in a certain section are rather sparse yet, in order to get a general sequence of climatic changes, the intervals between the dated intervals are assumed to represent periods of uniform deposition. Although it is accepted that this may reduce the accuracy of the chrono-stratigraphical division, it is still argued that it does not change the interpretation of the general pattern of climate changes during the lower part of the Holocene.

In the upper part of the Holocene, namely after *c.* 4 ka BP, the difference between the relative and absolute time scales diminishes as one progresses along the time dimension, and this, too, is within the boundaries of accuracy attainable for fixing the incidences of the climate changes based on proxy-data time series. We

must also take into consideration the fact that the correlation lines of iso-impact of climate change between various regions will cut iso-chronological lines, because of the difference in the pace of response of different systems and different regions. The correlation between climate changes and historical events should be considered as giving the general framework of concurrencies rather than that of the particular events. These general lines will be discussed in the following sections from the climatological, hydrological, archaeological and socio-economical aspects. When correlations between the proxy-data curve and archaeological events are given, the calibrated dates will be given, while the archaeological dates will refer to the BCE–CE calendar (see Table 1.1 for correlation of archaeological periods in the Levant).

1.4.1 The Neolithic period, *c.* 10 ka to *c.* 7 ka BP

During the Neolithic period, human society achieved some remarkable progress in its struggle for survival. In the first place, certain societies in the Middle East adopted agriculture and domesticated animals as the dominant strategy in their struggle for survival. This agriculture, as will be discussed below, was based on irrigation. By using irrigation human society became less dependent on the natural environment, where productiveness is dependent on the randomness of climate changes, especially in regions bordering the arid belts of the world. In the second place, pottery was invented. This enabled the storage and cooking of agricultural products and of those that were gathered or hunted. The third achievement was the creation of urban centers, which enabled the efforts of a large community of individuals to be concentrated for cooperative projects, such as a defensive wall against artificial or natural hazards, such as floods. In due time, such communal effect enabled diversion canals to be dug for the irrigation of land at a distance from the riverbed.

Were these achievements interconnected with climate change? I believe there is such an interconnection, though climate change acted more as an encouraging agent rather than the main basic cause. This basic driving force is the general evolution of intelligence of the bio-world in general, and of the human species in particular (Issar, 1995a,b). In this conceptual model (Issar, 1990), the interconnection lies in the fact that the change of climate in the Middle East, resulting from its location on the border of the desert, forced humans to develop modes of life that would enable survival under natural conditions that would have precluded survival without such changes. The most eminent support for this conceptual model can be found in the archeological excavations of Jericho in the Jordan Valley (Kenyon, 1957). In the period around 11 ka BP, the climate was more humid as a result of the Young Dryas, a colder climate, and one would expect human societies to expand into regions that had previously been deserts. When the climate at a later stage became warm and dry again, these

Table 1.1. *General archaeological time table*

Date	Egypt	Syria-Palestine	Mesopotamia	Anatolia
CE				
2000	Mamluk–Ottoman	Mamluk–Ottoman	Seljuk–Ottoman	Seljuk–Ottoman
1000	Early Arab period · · · · · · · · ·	Early Arab period · · · · · · · · ·	Early Arab period ·	
	Roman–Byzantine	Roman–Byzantine	Parthian–Sassanian	Roman–Byzantine
	Ptolemaic · · · · · · · · · · · · · ·	Persian–Hellenistic · · · · · · · ·	Persian–Hellenistic · · · · · · · · · · · · · · ·	Persian–Hellenistic
BCE	Late period	Iron Age II	Assyrian–Neo-Babylon	Iron Age
1000	· ·	Iron Age I · · · · · · · · · · · · · ·	· ·	· ·
	New Kingdom	Middle–Late Bronze Age	Old–Middle Babylonian	Middle–Late Bronze
2000	Middle Kingdom· · · · · · · · ·	Intermediate Bronze Age · · ·	Akkad–Ur III/Isin · · · · · · · · · · · · · · ·	· ·
	Old Kingdom	Early Bronze Age II/II	Early Dynastic I–III	Early Bronze Age
3000	· ·	Early Bronze Age I · · · · · · ·	Jemdet Nasr · · · · · · · · · · · · · · · · · · ·	· ·
	Archaic period	Mature Chalcolithic	Gawra (N)–Uruk (S)	Late Chalcolithic
4000	Pre-Dynastic period · · · · · · ·	· ·	· ·	· ·
		Early Chalcolithic	Ubaid (N & S)	Middle Chacolithic
5000	· ·	· ·	· ·	Early Chalcolithic · · · · ·
	Neolithic period	Pottery Neolithic A + B	Halaf (N)–Ubaid (S)	Ceramic Neolithic
6000	· ·	· ·	· ·	Early Ceramic
	Various	Pre-pottery Neolithic B	Hassuna–Samarra (north only)	Neolithic
7000	epi-Paleolithic cultures · · · ·	· ·	· ·	· ·
		Pre-pottery Neolithic A	Pre-pottery Neolithic (north only)	Aceramic Neolithic
	· ·	Natufian · · · · · · · · · · · · · ·	· ·	Epi-Paleolithic period
		Epi-Paleolithic: Kebaran		

From: Issar and Zohar, 2003.

societies concentrated near perennial rivers and springs, where they could put into more intensive practice the already existing rudimentary methods of planting and herding. The establishment of settlements near the perennial water sources enabled these societies to survive during drier periods and to develop and thrive during wetter periods, which later occurred.

1.4.1.a THE LOWER NEOLITHIC PERIOD, *c.* 10 ka TO 8 ka BP (PRE-POTTERY NEOLITHIC)

The lowest part of the isotope cross sections at Lake Van (Lemcke and Sturm, 1997) indicates that at *c.* 10 ka BP, the ratio of $\delta^{18}O/^{16}O$ was rather low, while the lower part of the oxygen composition of the marine curve (Luz, 1979) shows a heavy composition, which may suggest a rather cool and, therefore, humid climate. This may have facilitated the formation of sedentary communities organized in small villages during the beginning of the Neolithic period (Bar Yosef, 1986a; Clutton-Brock, 1978). Areas such as the Mediterranean zone, which were suitable for primitive agriculture, would rely upon cultivated plants and livestock. Semi-arid zones could provide hunting and grazing for domesticated animals. In any case, the transition from a hunter–gatherer to a farming society was not immediate. One can assume that the settling into commu-

nities only occurred after the acquisition of knowledge about food storage and the opportunities derived from cultivating plants.

On the basis of his own and other investigations in the Uvda Valley in the central part of Negev Desert of Israel, Avner (1998) concluded that agricultural settlements existed in this desert valley continuously from *c.* 10 ka to *c.* 4 ka BP. Bar Yosef (1986a,b) is of the opinion that, although communities of hunters and gatherers continued to exist in the arid regions such as the Negev mountains, the farming communities of the PPNA period expanded mainly towards the northern part of Israel and to the Jordan valley. During this period, the two main sites of settlements in Israel were Jericho and Nahal Oren.

Kenyon (1957), who excavated the site of ancient Jericho, maintained that the PPNA period was a period of floods, which again indicates a more humid climate. Jericho is situated in an arid region because it is shadowed from the rains by the Judean mountains. For this reason, I maintain that the development of agriculture at Jericho was based on irrigation (Issar, 1990). Such water could be derived from two sources: the floods coming from the mountains and the water from a big perennial spring, emerging from a regional aquifer, fed by the precipitation falling on the Judean mountains. It seems more logical that the farmers of ancient Jericho learnt to harness the water of the perennial spring before they found a way to use the water of the floods for irrigation. Moreover, the considerable storage of groundwater in the aquifer feeding the spring could mitigate the impact of a few years of drought. In any case, whether irrigation came from floods or from the spring, or from both sources, the fact that an agricultural society could survive in such arid conditions for about a thousand years suggests that the spring was perennial during this period and/or floods were abundant. Both suppositions lead to the conclusion that the climate was not too dry. Issar (1990) suggests that the floods became too strong towards the end of the period, and the disappearance of the PPNA people may have been caused by severe flooding, as evidenced by the nature of the layers of silt and gravel that cover that ruins of the PPNA culture.

Nahal Oren, by comparison, is situated in the more humid part of Israel, on the fringe of Mount Carmel near the Mediterranean coast, and enjoys ample rains during the winter season while, during the summer months, the natural forest of the Mount Carmel, and its fauna, could supply ample food during humid periods. Even so, a few years of continuous drought could have forced the settlers to abandon the place.

Therefore, one has to conclude that the climate was indeed rather cold and humid and it is difficult to agree with the conclusions of Horowitz (1973, 1980), who maintains that the pollen assemblage in Lake Hula (Huleh) indicates higher temperatures during this period of time and thus concludes that this period was drier than that of the PPNB. The same difficulty arises with the suggestion of Van Zeist (1969) that the Near East suffered from a rise of temperature during the PPNA, and that the climate during the PPNA was too dry to allow cultivation of barley and wheat; consequently the grains that have been found in the Near East must have been imported. Also Butzer (1978) concluded that temperatures rose during the PPNA and precipitation was less than that today. A marked improvement in precipitation only occurred towards the beginning of the PPNB. Wreschner (1977) found little material from the PPNA in the coastal region. The sites that were found lie high above present sea level. He explains this as a consequence of the Flandrian transgression of the sea, which forced the inhabitants to look for high ground for their living sites. Based on the evidence from Jericho, I would suggest that this period may have been characterized by severe storms, high groundwater tables and strong floods, which caused the people to seek higher ground for their settlements. Indeed Sakaguchi (1987) reports that a lake existed in the Palmyra district between 10 ka and c. 8 ka BP.

The disagreement between these assessments may be explained by changes in climate during this period, starting with a cold climate, continuing as warm and dry, and becoming again colder and more humid, reaching a maximum, towards the period's end. As vegetation tends to change rather slowly, evidence based on pollen will indicate a dry climate, while evidence from iso-

topes, hydrological systems and human habitat, all of which respond rather quickly, will indicate that the period was cold and humid.

The later gradual enrichment of the deposits of Lake Van by heavy isotopes of oxygen and, at the same time, the depletion in the oxygen isotope ratios of the biomarine deposits suggest the warming up of the climate towards the end of the PPNA. There is also evidence that the desert areas were settled at this time, which suggests a climate warm enough to propel the monsoon-type rains northward, causing summer rains over the deserts of the Negev and Sinai.

Bar Yosef (1986b) suggests that the climate during the PPNB favored hunters and gatherers and attracted them to occupy more desert areas (the Negev and northern Sinai). The economy of the sites in the more humid areas was based on legumes and cereal cultivation, together with hunting and herding. In the more arid areas, the inhabitants probably lived in the sites during the winter, autumn and spring, with an economy based on hunting and gathering. A remarkable number of sites were found in the central part of the Negev (Gopher, 1981). According to the pollen spectrum found in one of these sites (Sede Divshon), Horowitz (1977) came to the conclusion that this period was more humid than today and enabled agriculture even in the more desert area. A pattern of dense PPNB settlement is found in the central Sharon and southern plain (Wreschner, 1977), which supports a climate during most of this period that was humid enough to sustain socio-economic systems based on agriculture.

Sanlaville (1989), who investigated the sediments of the Persian Gulf, observed a transgression from 9 ka to 6 ka BP, expressed by a rapid progress of the Persian Gulf shore towards the north, which also suggests a warm period.

Toward 8 ka BP, the oxygen isotope composition of the continental layers at the bottom of Lake Van became lighter and the isotopic composition of the planktonic foraminifers at the bottom of the eastern Mediterranean showed a heavier trend. At the same time, the level of the Dead Sea was rather high. These signals point towards a cooler period, and most probably a more humid climate. From palynological analysis, Erinc (1978) concludes that the climate in southwestern Anatolia was markedly cooler and moister at c. 8.5 ka BP than it is now. This stage of relatively more humid conditions was followed by an extremely dry phase at c. 7 ka BP.

1.4.1.b THE MIDDLE AND UPPER NEOLITHIC PERIOD, c. 8 ka TO c. 7 ka BP (POTTERY NEOLITHIC)

The heavier composition of oxygen isotopes in the continental environments, the lighter composition of oxygen isotopes in the marine environment and the lower level of the Dead Sea, all betoken a warmer and presumably drier climate during the Pottery Neolithic (PN) period.

The occurrence of deposits of greenish-gray laminated sediments led Neev and Emery (1967) to suggest a humid period from the beginning of the Holocene until 7 ka BP. Thereafter, the climate became drier, reaching a peak between 6.5 ka and 5.5 ka BP. Neev and Hall (1977) then revised their study of the depositional processes of the Dead Sea. The new study was based on new U/Th and ^{14}C dates and updated stratigraphical information. They concluded that the period when the paleo Dead Sea (Lake Lisan) extended over the Jordan Valley was followed by a pluvial period lasting 3000 years, from 10 ka until about 7 ka BP, while the dry period, evidenced by extensive rock salt deposition, occurred between 7 ka and 5 ka BP.

Very few PN sites were found in the northern part of the Negev and southern Sinai although six sites were discovered in the southern Negev. Most of the PN sites in Israel are found in the Jordan Valley and the coastal plain. Bar Yosef (1986b) explained the scarcity of sites, not as a consequence of climatic conditions but as a consequence of social and economic changes (an organization of "tribal kingdoms"). However, Horowitz (1973, 1980) uses palynological data to conclude that there was an increase in temperature. He also maintains that there was a rise in the sea level by 1–2 m as a result of melting ice sheets. There was an increase in precipitation even in the southern parts of Israel at about the same time. The herbaceous cover in the south, resulting from quite a heavy level of precipitation, changed the economic system and directed the emphasis towards agriculture and pastoralism. Crown (1972) and Issar et al. (1992) explained the increased humidity in this area as being caused by increased precipitation owing to the migration of the monsoon belt northward. It is suggested that from about 7.5 ka BP there was a reversal in the relationship between the anticyclones of the Azores and northeast Europe, which had an effect on the climate of the southern part of the Near East. Up to about 6.5 ka BP, the southern part of the Near East came under the influence of the trade winds and thus the summer monsoon rains. As a result, the PNB climate in the southern part appears to have been both warmer and moister than that today, while in the northern part, it was warmer and drier.

In general, most of the PN sites (8 ka to 6.2 ka BP) are covered by alluvial deposits with large stones, indicating severe flooding. The Neolithic sites were located close to water resources and in low areas, which required a protective system against flood and mud flow damage. As Bar Yosef (1986b) noted, the response of the inhabitants of Jericho to the floods and sheet wash was "to build a wall and then, where necessary, dig a ditch". Knowing the Neolithic inhabitants as a peaceful society (lacking social aggression) before the eigth millennium BP, it is an alternative explanation for the Neolithic walls of Jericho.

There was an increase in oak and the appearance of pistacio in the vegetation in the regions bordering the eastern Mediterranean Sea after c. 10 ka BP (Post Glacial Period). This indicates both a rise in temperature and an increase in humidity as drought caused by a rise in temperature alone at lower elevations would have been a limiting factor for oaks (even though oak is more tolerant than conifers of dryness). The appearance of oak is an indication of an accompanying increase in precipitation. The climate change favored an extension of the forest, while steppe plants mostly disappeared. The eastern parts of the Near East zones, like the Ghab Valley in eastern Mesopotamia, seem to show a similar pattern of expansion of the forest (especially oaks) to that found in Greece and Italy. Information from more easterly zones like northwest Iran (Lake Zeribar) point to a change in climatic conditions, but the oak forest reached an optimum only at about 5.5 ka BP (Bottema, 1978).

Wright (1976) concluded from pollen records in lake sediments that steppe vegetation changed to open woodland or to forest at c. 11 ka BP. In some areas, like Lake Zeribar, the transition started as late as 6 ka ago whereas in others such as Tenagi Philippon, Macedonia, the transition started as early as 14 ka BP. Wright also claims that the vegetation change "perhaps reflects variable responses of local areas to increases in precipitation or temperature or both, as well as possible delays in the migration of trees from Pleistocene refuges". Van Zeist and Bottema (1982) came to the same conclusion concerning the heterogeneity of climate from one zone to another within the same climatic belt.

Depletion in ^{18}O values in the carbonate of land snail shells from 9 ka to 7.3 ka and from 6.5 ka to 6 ka BP (Chalcolithic period) was related by Goodfriend (1991) to changes that occurred in circulation patterns during more humid periods. During those periods, the rain entered the Negev area from northeastern Europe through the Mediterranean Sea to northeastern Africa. The depletion in ^{18}O values results from intensive evaporation over the Mediterranean Sea or from a continental effect along the northern coast of Africa. Goodfriend (1991) suggests that the changes in ^{18}O represent changes in the isotopic composition of rainfall in the Negev, rather than temperature fluctuations. The analysis of ^{13}C in organic matter in the early Holocene land snails also supports the above findings, by showing approximately twice the rainfall in the northern Negev during the early Holocene compared with that at present (Magaritz and Goodfriend, 1987). In another study, Goodfriend (1990) found that there was a shift of c. 20 km to the south of the transition zone of 150 mm isohyets during the period between 6.5 ka and 3 ka BP. On the basis of these data, the Negev zone would have had more rain until 3 ka BP. This time scale, based on the snail data, was of a longer humid period than is suggested by lake and sea sediments. I would suggest that these data are considered as a general indication of climate, and in some periods they may provide an alternative scenario to that suggested in this book.

On the whole, it can be concluded that the Neolithic period, extending from the PPN and PN periods, was characterized by climatic fluctuations. The general trend seems to have been humid during most of the period and dry towards its end.

1.4.2 The Chalcolithic period, *c*. 7 ka to *c*. 5 ka BP

The Neolithic period ended towards the end of the seventh millennium BP and was followed by a culture characterized by a new innovation, namely copper production. The Chalcolithic culture arrived in the Middle East at about 7 ka BP. The period between the Neolithic and Chalcolithic is demarcated by a gap in settlement that might have been caused by an extreme phase of the warmer and drier climate, which reached its maximum around 7 ka BP but might have extended later to influence the pattern of settlements a few centuries later.

A marked depletion of the ^{18}O composition of Lake Van can be observed in sediments from *c*. 6.5 ka BP (Lemcke and Sturm, 1997) and continued to about 6 ka BP. Degens *et al.* (1984) observed a strong rise in the level of Lake Van during this period. The low ^{18}O values of planktonic foraminifers from deep-sea cores in the eastern Mediterranean (Luz, 1979; Luz and Perelis-Grossowicz, 1980) suggest the inflow of heavier ocean water. Although no evidence was found in the caves of Mount Sodom to indicate that the level of the Dead Sea was high during this period, I believe that the still higher level of the lake during the lower Bronze Age, which followed, obliterated any such evidence. The evidence for a more humid climate will be discussed below.

An important settlement of the Chalcolithic period was excavated at Tel el-Ghasul in the eastern Jordan Valley near Jericho (Hennessy, 1982). The archaeological remains in this site show that it was inhabited by a society that reached a rather sophisticated cultural level, building a shrine and decorating its walls with mythological murals. This culture is referred to as Ghasulian, and it is believed to have started sometime during the third quarter of the sixth millennium BP (Ussishkin, 1986).

The origin of the people of the Chalcolithic period is still unclear. The main stream in Middle Eastern archaeology believes that part of the population of the Chalcolithic culture was local and it absorbed a new wave of people with their new culture. Some of the researchers note that there was a migration of people from the "north" (Mellaart, 1966; Govrin, 1991). Gophna (1983) also claimed that the Chalcolithic population entered the Levant from the north and brought with it a very developed and organized culture. Ussishkin (1986) suggested that, according to archaeological remains and skeletons' structure, the Chalcolithic cultural bearers originated in the Caucasian Mountains of East Anatolia or the mountainous areas of Armenia. Other archaeologists accept the idea that the new Chalcolithic immigrants came from an unidentified "east" (Elliot, 1978), which could be Mesopotamia. Yet, not all archaeologists agree with the "north or east theory"; for example, Gonen (1992) suggested that natural and internal processes were responsible for the change from the small isolated local communities of the Neolithic period to the new social and economic system of the Chalcolithic period. The change in culture was in a response to the need to find a solution for the changing condition of the society and to ensure its continuation and succession.

Whichever theory is correct, one still has to explain the reasons for the sudden rise, after several hundred years of decline, in the number of settlements, and the appearance of a new form of culture equipped with a new technology, for which there is no sign of a gradual evolution, as is the case in the invention of pottery. Issar (1990) favors the theory of the immigration of people from the north and explains it by a strong climatic change from warm and dry to cold and humid. This made the high plateaus of Anatolia, Iran and maybe even Central Asia less habitable, and it simultaneously caused the plains of the Middle East to flourish. This, as archaeological evidence shows, gave rise to an incredible increase in population density: a rise in the total numbers as well as an increase in community sizes and the range and rate of productivity in many economic areas. New activities included ceramic, metal, ivory and basalt industries. Levy (1986) suggested that the Chalcolithic economy indicated a development of production beyond the domestic circle and based on an increase in socio-economic complexity, which involved the development of social ranking and hierarchies. Another characteristic feature of the Chalcolithic culture was the emergence of distinct regional cultures, with a high level of adaptation to the local environment (Levy, 1986).

The population of the Chalcolithic period settled in Israel in planned farming communities in the Jordan valley, the coastal plain, the Judean desert and the northern and eastern Negev (Ussishkin, 1986). Sedentary village life was established during the Neolithic period, but the Chalcolithic cultural communities were larger and more advanced farming villages. These villages later became the pattern of the "modern" village in the Middle East.

These settlements expanded into the Negev. Their remains are found all over the Beer-Sheva plain, reaching the Arad area to the east and Nahal Besor to the west. They spread into the arid Arava valley, from Ein Yahav in its northern part to Timna in the south. Most of the settlements were located on the tops of low hills close to river valleys (Cohen, 1986, 1989). Archaeological remains show that many settlements also thrived in the valleys of the southern mountains, practicing agriculture in areas that today get less than 100 mm of rain per year (Avner, 1998). Cohen (1989) maintains that the settlements in the Negev mountains were temporary and semi-nomadic, based on pastoral grazing and transportation of copper from the Feiran area and the Timna valley.

The discovery of mining and smelting sites in the Feiran area, the Timna valley and Eilat area suggests the importance of mining and special production activities. Avner (1998) attributes this flourishing of the desert to a more humid climate. He further suggests that the source of the rain was monsoonal. Yet, this explanation makes it difficult for him to find a climatic reason for the fact that this valley continued to flourish during the Early Bronze Age (EB); he explains this anomaly anthropogenetically, by the adaptation of these societies to desert conditions. This explanation may not be necessary, however, if one considers the EB to have been cold and humid, as I suggest.

The settlers in the Beer-Sheva plain built their dwellings underground, digging into the loess soil in the escarpment overlooking the riverbed (Perrot, 1968). The people most probably received their water supply from shallow wells located in the riverbed. It is even possible that the river flowed during most of the year. There are many indications that the people cultivated fields along the riverbeds. Diversion dams were also used in order to bring water from the river to the fields (Alon, 1988).

Consequently, one cannot avoid the conclusion that, indeed, the Chalcolithic period was one of economic and cultural prosperity, and the most logical reason for this is a climatic change that brought more precipitation to the semi-arid Middle East and enabled agriculture to spread into the desert area. As noted, the isotopic evidence supports this conclusion.

The magnificent Chalcolithic culture, with its artistic tradition and technical knowledge, lasted for about 1300 years and suddenly disappeared towards 5 ka BP. All the sites were abandoned without any signs of violence. Archaeologists explain the disappearance of the Chalcolithic culture in various ways. Hennessy (1982) suggests that they had to leave their settlements because of a migration of a new wave of people – the ones who established the EB culture. But archaeological remains do not reveal any exchange of cultures in the same sites. Others looked for a circumstantial connection between the disappearance of the Chalcolithic population and the expansion of the first Egyptian Kingdom of Naarmer at about 5 ka BP. However, there are no indications of violence involved in the abandonment of the settlements. Levy (1998) suggests that a drier climate caused the collapse of this culture and I would support this explanation based, once more, on the isotopic data of Lake Van, the Sea of Galilee and Soreq Cave. Also, there appears to be an increase in the level of the Mediterranean Sea at this time. Yet there are some difficulties with this hypothesis, because it is known that the settlements in the area of Beer-Sheva and the Judean desert were the last to be abandoned (Gonen, 1992). Perhaps the type of settlement that was established in the south can explain this. The people of the south maintained either a pastoral way of life or agricultural settlements along the riverbeds (Govrin, 1991). Both societies experienced semi-arid conditions from time to time. Therefore, the conditions led them to develop a way of life involving desert agriculture irrigated by floodwater and the technology of shallow wells. This might have given them a sufficient water supply during less rainy years. It is

reasonable to think that a severe crisis in the north caused the rapid collapse of the settlements in that area, while the impact of the crisis on the population of the more arid zone, who exploited their knowledge and life style, was less severe, at least in the beginning. A further explanation, with a similar basis, is that the warming up of the climate increased the incidence of monsoon-type rains and this helped to support vegetation suitable for forage.

In the general framework of these paleo-environmental and socio-economic scenarios, one can explain the observations made by Tsoar and Goodfriend (1994) of a dense population existing on the sand dunes of the northeastern Sinai bordering the Negev at *c.* 4100 cal. BC. According to these authors, the higher silt content of the sands indicates a higher rate of precipitation. They suggest that the activation of the dunes was a function of overgrazing and trampling. An examination of the precipitation curve from Soreq Cave (Fig. 1.3) shows that precipitation reached a peak during this periods, but soon afterwards the climate started to deteriorate. I would suggest that the activation of the dunes was more a function of overgrazing during the aridization phase, which immediately followed the peak (and which can be seen in the Lake Van humidity record in Fig. 1.3).

Sanlaville (1989) records a maximum level of the Persian Gulf at *c.* 5.5 ka BP. Yet, as Sanlaville states, the dates on which this curve is based (Sanlaville, 1989, p. 19) show the high sea level to be around 5 ka BP, which corresponds well with the upper Chalcolithic warm period.

1.4.3 The Early Bronze Age *c.* 5 ka to *c.* 4 ka BP

All environmental data show that a major change some time after 5 ka BP brought a cooler and more humid climate to the Middle East. The level of the Mediterranean Sea declined, while the level of the Dead Sea rose, and oxygen isotope composition in lake deposits and speleothemes became lighter (although the change observed in the sediments of the Sea of Galilee appears to come later and is believed to reflect rather the small number of dated samples and thus the imprecision of the timing of the changes). Rosen (1986), investigating the alluvial deposits of Nahal Lachish, concluded that the deposits of the Chalcolithic period and the EB indicated a climate that was moister than today. Massive alluviation, indicating a more humid climate, was observed in other riverbeds, such as Nahal Beer-Sheva, Nahal Shiqma and Nahal Adorayim, in the southern part of Israel (Goldberg and Rosen, 1987). Consequently, more or less parallel to the time during which indications of proxy-data show a more humid climate, the archaeologists place the beginning of the EB. This brought major developments in agricultural technology, together with an enlargement of international trade, contributing to the emergence of the walled city, which achieved a central role as a religious and economic center.

There are some questions concerning the beginning of the EB. It is not yet clear whether it started immediately after the end of the Chalcolithic period, if there was an overlap between those two cultures or if a gap in time existed between the end of the Chalcolithic period and the beginning of EB I. Until the middle 1970s, the transition from the Chalcolithic period to EB I was explained by intrusion of foreign societies (de Vaux, 1971; Kenyon, 1979; Lapp, 1970). At the end of the 1990s, there are more and more claims that the transition from the Chalcolithic period to EB I was a process of local evolution (Levy, 1986; Schaub, 1982).

The debate whether this culture developed locally or was brought in by immigrants from the north is important with regard to the question of whether this cold period was strong enough to drive people from areas becoming less habitable because of the cold climate, towards the more hospitable south, for example the people of the central plateau of Asia. Although this issue has not been decided, it is interesting to view the relevant evidence.

Archaeologists who entertain a foreign origin for the EB I culture have noted that the urban life style had the character of a new culture. Kenyon (1979) described a new culture brought in by the Proto Urban people. She suggested that three groups of people (A, B and C) entered Canaan from the north and brought with them different types of craft. These Proto Urban groups existed side by side in different areas in the Levant, and all of them were responsible for the development of urban life. Lapp (1970) and de Vaux (1971) suggest scenarios differing slightly from that of Kenyon but agree that the B culture peoples came from the north with a new tradition of architecture and an urban life style.

Some difficulties in accepting the migration theories arise from the fact that there are no parallels to the emerging EB culture outside Canaan (Ben Tor, 1992; Levy, 1986; Schaub, 1982). Also, there is no proof for a route of migration to this place during that period of time. Moreover, on-going archaeological research has shown a connection between the cultural material of the Chalcolithic period and that of EB I. Ben Tor (1992) noted that ceramics from the EB, especially those belonging to the first stage of the EB I, were not totally different from those of the Chalcolithic period. In fact, there seems to be a sequence in traditions, which might also support the possibility of a continuation in the population.

Considering new evidence collected since the early 1980s, there is a tendency to explain the social processes leading to urbanization as an evolutionary conceptual rather than an intrusion: diffusion model (Ben Tor, 1992; Amiran and Kochavi, 1985). The evolutionary theories emphasize the continuation of local elements rather than "importation" from the outside, and the adaptation of the local culture to changing environments. Amiran (1985) emphasized the emergence of walled cities, which seem to have been established by the same type of population as in the preceding periods. Amiran saw continuity in population throughout the

early phases of the EB and concluded that there was an unbroken development from the village community to the urban society. But Amiran is cautious about concluding that urbanism was solely a local process. Hennessy (1982) also noted a mixture of foreign and local cultures in developing the so-called urban culture. Schaub (1982) concluded that the transition from the Chalcolithic period to EB I in Israel was locally oriented, as the new cultural material appears only gradually, rather than suddenly.

The size of the settlements of the EB I was almost the same as that of the settlements during the Chalcolithic period, but the density of inhabitants in most was increased. This is an indication of an increase in population (Ben Tor, 1992). However, not all the settlements during EB I are defined as cities. Some were "small cities", such as Megiddo, Lachish and Jericho. Others were considered "big cities", such as Yarmut, Gezer, Afek and Beit Yerach. Agricultural villages were located alongside the cities.

There is a transition of EB settlements to the mountainous areas, the foothills and the valleys of Jezreel. New locations for sites involved a transition to a Mediterranean method of agriculture. Ben Tor (1992) and Broshi and Gophna (1984) claimed that the people of the EB mainly preferred areas in which the annual amount of rainfall was more than 300 mm. More than 600 sites from EB I and II were discovered in the Negev highlands and Uvda valley (Avner, 1998), which would support the claim that it was a humid period.

Not all the EB settlements of the Negev were walled and most of them were located on hills, near valleys, that had permanent water resources (Cohen, 1989). Cohen claimed that the population growth of the Negev highlands was caused by the movement of settlers who could not adapt themselves to the growing urbanization of life style in the northern parts of Israel. The economy of the EB settlements in the Negev was based on agriculture, pastoralism and hunting.

The archaeologists divide the EB into three parts, namely EB I, EB II and EB III. However, there is not a major change in the basic cultural characteristics during these times. Most of the EB I settlements became more urbanized during the EB II and a type of a "planned city" appeared, parallel to a transition from settlements with no walls to fortified cities (Ben Tor, 1986, 1992; Broshi and Gophna, 1984). Thus, the seeds of urbanism, which were planted during EB I, flourished. Walled cities and agricultural villages were located side by side, and economic and social connections were developed between them.

At *c.* 4.6 ka BP, the city of Arad, located in the eastern part of the Beer-Sheva plain on the border of the desert, was deserted, most probably because of worsening climate (Amiran, 1986; Amiran *et al.,* 1980a,b). When I investigated the water supply system of ancient Arad, I found that the main supply of the EB Canaanite city was based on a deep well touching a perched local groundwater table. Its recharge area was limited to the area surrounding the city. Consequently, once the climate started to become drier, replenishment diminished and the well dried up, leaving the inhabitants of the city without a perennial water supply (A. Issar, unpublished report). Whether the desertion of Arad was the first sign of the major dry spell characterizing the end of the EB is difficult to say. A general trend towards drying up of the region is indicated by several pieces of evidence: the city was not resettled during the MB; the isotopic data shown in Fig. 1.2; the paleo-precipitation curve calculated for Jerusalem from isotope data in the speleothemes of Soreq Cave; and the decline of the level the Dead Sea, (Fig. 1.3). Samples of ancient tamarisk wood found along ancient shorelines of the Dead Sea, which penetrated the caves of the salt plug of Mount Sodom (Frumkin *et al.*, 1991), dated to 3780 years BC. These were analyzed for their ^{13}C content and gave values similar to that of present-day trees in this region, namely for trees growing in an arid environment, By comparison trees dating from the Roman period, *c.* 2 ka BP, showed a marked depletion of ^{13}C, indicating a cooler and more humid period. Additional areas of desertion were observed in other more arid parts of the Middle East (Richard, 1980, 1987). Most of these sites were dependent, like Arad, on local springs. In general, a survey of the archaeological reports shows a desertion of the settlements in the Negev and Sinai during EB III. One can conclude that a warming-up phase had started some time towards the end of the third millennium BC, and, as a result, a gradual decline in the average annual precipitation was first evidenced in the areas closer to the desert. Settlements that were in the vicinity of perennial springs, feeding from a regional aquifer (such as Jericho, Megiddo, Beit Shean), close to the Sea of Galilee (Beit Yerach) or in the more humid parts of the country like Ai and Lachish continued to prosper. It is possible that many of the inhabitants is of the abandoned settlements of the EB II in the drying-up regions resettled in the more humid areas or at sites that were located near big springs, such as Jericho.

As time went on, the trend towards warming strengthened and a general trend of decline of cities, with a transition to a nomadic and semi-nomadic society towards the end of EB IV, namely towards *c.* 2.3 BC (or *c.* 3.8 BP uncalibrated) is reported by all archaeologists. The exact time period of EB IV is a matter of debate between the archaeologists (Albright, 1962, 1965; Amiran and Kochavi, 1985; Ben Tor, 1992; Dever, 1973; Gophna, 1992; Lapp, 1970; Mazar, 1986; Wright, 1938). As this book centers on the general trends of climate change and their impact on the hydrological and socio-economical systems, and, of course, different water systems and various societies will react at a different pace to the worsening of the climatic conditions, the main conclusion that one can draw from the archaeological reports is that at the turn of the third to the second millennium, the desertion of cities reached a climax. At the end of this stage, urban centers, which had existed for several hundreds of years, had all disappeared, being abandoned without any sign of destruction. The demise of the cities has been observed in archaeological excavations all over the Middle East (Butzer, 1958; Harlan, 1982; Rosen, 1997; Weiss *et al.*, 1993).

In Canaan, EB IV is difficult to recognize, even in the cities located near perennial springs, like Hazor, Megiddo, Beit Shean, Jericho, Lachish, and Tel Beit Resisim (Dever, 1980, 1985a), most probably because of the growing poverty of the surrounding agricultural communities. With the collapse of the cities in the northern more humid part of Canaan, the center of settlement was shifted to the semi-arid and arid marginal areas such as Trans Jordan, the Jordan Valley, the Negev and Sinai. This is explained by the archaeologists as a result of changes in the socio-economic structure of the local population, as the demise of the cities of the north forced the people of these cities to move to the marginal areas and become pastoral nomads (Dever, 1985a; Richard, 1980, 1987). "Pastoral nomadism" according to Johnson (1969) is distinguished from a sedentary way of life by the high degree of mobility and the adjustment to seasonal availability of pasture and water for animal husbandry, which is the primary economic basis. Finkelstein and Prevolotsky (1989) recognized the cultural, political and social relations between nomads and urban populations. Neither group was totally independent. Therefore, the collapse of urban society caused pastoral society to look for a new way of supplying its needs for sustenance. The change in economy of the pastoral society towards one of grazing and agriculture gave rise to a sedentarizing process among the nomads in the marginal areas.

The reasons for this regional devastation are a matter of argument between the archaeologists, most of whom do not accept climate change as a reason. Some archaeologists connect the abandonment of cities with the migration of the Amorite, a semi-nomadic people who overran the EB III cities, causing de-urbanization (Albright, 1926; Kenyon *et al.*, 1971; Cohen, 1983). However, they fail to consider whether there was an environmental reason for the Amorite migration. The Amorite hypothesis is based on Sumerian texts dealing with the invasion of tribes from the east into Mesopotamia. Such an invasion should have left signs of destruction, but the archaeological remains indicate a continuation of the cultural material of the EB IV with the EB III tradition (Richard, 1980, 1987), which makes it more reasonable to attribute the cultural material to the same population rather than to a new wave of immigrants. The same argument applies to the rejection of the "invasion theory" of Trans-Caucasus tribes, who were supposed to have come from Central Asia (Kochavi, 1969; Lapp, 1970). Mazar (1968, 1986) suggested another explanation for the disappearance of the cities. He claimed that the Egyptian campaign caused a dislocation of people and ruined cities, and this was the cause for the decline of the political regime in Israel. This kind of disorder left the country open to the Amorites, who came with the Acadian troops and replaced the weakening Egyptian troops. However, archaeological remains do not support any destruction by military forces. Moreover, there is no textual evidence of Egyptian military activity. Ben Tor (1986) noted that one of the explanations for the de-urbanization could have been the competition between the cities themselves. This hypothesis does not have any archaeological evidence.

Richard (1980, 1987) was one of the few archaeologists who came to the conclusion that it was a climate change that had such a destructive impact on the socio-economic systems of the Middle East. He concluded that, although amalgamation of a complex array of socio-economic and political factors must have occurred to terminate the EB urban life, probably the major factor was a shift in climate to drier conditions. This ecologically significant shift caused, either by itself or combined with an already weakened economy, the abandonment of sites. Presumably, the climatic shift was substantial for otherwise one would expect cultural adaptation to the new condition rather than total abandonment of the sites (Richard, 1980, p. 25).

As already mentioned, the environmental proxy-data presented in Figs. 1.2 and 1.3 support this explanation. Moreover, in archaeological surveys where the archaeologists examined the nature of the deposits as well as the cultural remains, they found evidence that, indeed, the lower part of the EB was humid, while the upper part became dry. Ritter-Kaplan (1984), in her report on the excavation of the Exhibition Gardens in Tel-Aviv, described the different characteristics of the soil stratification at the site. She found that a black clayey layer laid down in a swamp and containing an abundance of oak pollen was deposited during the EB I–III age. Overlying this layer, she found a grayish sand layer, almost devoid of pollen, overlying the remains of the EB IV or MB I culture. She interpreted the change from clay to sand as an indication of a change to a drier climate and described this period as that of "the crisis of the aridity".

The sand, as already discussed, came from the Nile, and the increase in its supply was most probably a function of an increase of the precipitation on the Ethiopian highlands and possibly from the erosion of the delta through abrasion by the rising sea level. Sivan (1982), investigating the history of deposition in the Haifa Bay, found that this bay was dramatically filled in with sand, resulting in the shoreline advancing several kilometers since the MB, c. 4 ka BP; this is additional evidence for a major climate change.

The pollen assemblage in the sediments of the Sea of Galilee shows a remarkable reduction of the pollen of the Mediterranean natural forest, especially of oak and pistachio, and a parallel increase of olive pollen starting c. 5.5 ka BP, and the reversal of this trend starting c. 4.5 ka BP (Baruch, 1986; Stiller et al., 1983–84). I interpret the first stage as a result of a more humid climate, which made it economic for the people of the semi-arid part of the drainage area of the Sea of Galilee to cut down the natural forest and plant olives. When climatic conditions started to become warmer and drier, the olive harvests declined to a stage when they

had to be abandoned, and the natural vegetation rejuvenated. Afterwards, one finds a similar trend during the shift from the Roman to the Moslem period.

Quercus and *Pinus* species grew at Beer-Sheva and Arad (northern Negev) at 2.8 ka BP. Later, and to the present day, these trees completely disappeared and were replaced by *Acacia* sp. and by tamarisk, indicating an arid to semi-arid habitat (Liphschitz and Waisel, 1974).

As already mentioned, this climatic crisis affected the whole Middle East. Neuman and Parpola (1987) found documentary proof of aridization in a reduction of the water level of the Tigris–Euphrates and an increase in salinization (from c. 4.3 ka to 3.9 ka BP). Crown (1972) found indications that the climate in Iraq after 4.5 ka BP was drier than the markedly wet period of 5.5 ka to 4.5 ka BP. About 4.3 ka BP, the rise in temperature caused severe droughts and crop failures. Lakes in the Zagros mountains dried out between 4.5 ka and 4.0 ka BP (Wright, 1966) and a significant reduction in arboreal pollen was found in Lake Zeribar (Van Zeist and Bottema, 1977). Weiss et al. (1993) analyzed soil samples from Tel Lailan in northern Mesopotamia and also concluded that a deterioration in the climate had caused the geo-political crisis which led to the demise of the Acadian empire at the end of the third millennium BC. The same conclusion – that a severe climate change was the reason for the collapse of the socio-economic systems of the Levant – was reached by most of the participants in the NATO workshop in 1994 (Dalfes et al., 1997).

All this evidence tends to refute the accusation that the ancient inhabitants of this region caused the salinization of their soils. This claim was based mainly on data from deciphered clay tablets from Sumerian archives (Jacobsen, 1957–58, 1960; Jacobsen and Adams, 1958). The data demonstrate that many long irrigation and (probably also) drainage channels were dug by the inhabitants of that area between 2300 and 1800 BC. In this same period, the ratio of barley to wheat was constantly rising in offerings and taxes delivered to the temples. As barley is more tolerant of soil salinity than wheat, the archaeologists incorrectly concluded that excessive irrigation had caused salinization. There was an attempt to refute these accusations by demonstrating, from similar archive texts, that the Sumerians were aware of the danger and had taken some preventative drainage measures (Pollock, 1999).

Sanlaville (1989) observed that the delta of the Mesopotamian rivers in the Persian Gulf rapidly developed at c. 4.5 ka BP. It is suggested that this could be correlated with the cold and humid climate of the EB.

To sum up, environmental and archaeological evidence indicates that the climate was wet during the lower part of the EB, presumably as the result of a cold climate phase. A dryer phase started in the middle part of the EB and reached its climax towards the end of the fifth millennium. The deterioration of climate caused the collapse of the urban socio-economic system of the Middle

East as city after city was deserted. The widespread nature of this phenomenon leads one to the conclusion that it must be connected with a global climate change.

1.4.4 The Middle Bronze Age, *c.* 4 ka to 3.5 ka BP

As already noted, archaeological data attest to a drying of the climate towards 4 ka BP. At the same time, there are indications of a warmer and drier climate from the proxy-data time series. The ^{18}O and ^{14}C composition of the Sea of Galilee sediments and speleotheme caves increased; there was a rise in the level of the Mediterranean Sea while the level of the Dead Sea fell; and there was a decrease in the number of plankton foraminifers in the eastern part of the Mediterranean and the isotope composition of the benthos and pelagic foraminifers became heavier, apparently because of a greater flux of fresh and heavy Nile water.

This climate change also led to a transition into a new culture, namely that of the MB. Although archaeologists differ in their opinions as to where exactly to place the border line between the EB and MB, all agree that there was a transition from one type of culture to another at *c.* 4 ka BP, separated by a period of a few hundred years (*c.* 4.3 ka to 4.0 ka BP), when a massive collapse of the urban centers of the EB took place and there was a return to a pastoral and nomadic way of life. The archaeologist Dever (1987) named this period the "dark age" of the uppermost EB. On the basis of the paleo-environmental data, this period can also be named the "dark age" of most of the lower MB, before the beginning of the MB renaissance period.

As this period started, some time after 4 ka BP, a major socio-economic change took place in the southern Levant. Most of the abandoned urban EB sites were recovered and resettled. Moreover, many new sites were established all over the region. At the same time, the pastoral villages that had survived the "dark age" of EB IV in the more arid parts of Trans-Jordan, the Negev and Sinai were abandoned in favor of the central centers. The urban growth also indicated a more complex social organization, with new relationships between the urban centers. Concomitantly, many large cities started to become fortified, massive fortification continuing during MB II (3.8 ka to 3.65 ka BP) and MB III (3.65 ka to 3.5 ka BP). It seems that the main reason for the massive fortification was mutual hostility between local city-states, as well as the threat of outside invasion (Dever, 1987).

Although the MB is considered as an urban period, and the walled cities characterized this era, the peasants and agriculture systems were still very important components of the social structure. The economy was based on agriculture, with some local crafts, domestic industry and trade in luxury items. The political organization of the MB I still lacked centralization or the domination of one city. It seems that the tribal–patriarchal political system continued to dominate. The political urban regime, based on one ruler in each urban center, crystallized only during MB II (Mazar, 1967). So, while the MB II city was characterized by social complexity and integration, it was still far from the high level of socially developed city that already characterized the states of Mesopotamia and Egypt before 5 ka BP (Dever, 1987).

The urban cultures of the MB II and MB III were much more prosperous than that of the MB I. However, the roots of urbanization are to be found in the initial phase of the MB I. After the "dark age" of EB IV, the following period became one of a "cultural explosion". Albright (1949, 1973), Mazar (1967), Dever (1976) and Aharoni (1978a,b) all claim that the MB I culture could not have developed from local culture. The most accepted theory concerning the origin of the people of MB I suggests that their origin was in the areas of Syria and Mesopotamia (the Amorite tribes). These peoples came from urbanized regions and brought their rather new material culture with them to the Levant. Slowly, the local population, which had been assimilated into the Amorites, took on the material culture of their assimilators. Based on the names of the Canaanite rulers, their origin was Western-Semitic. But there are also Hurian elements from the "Land of Hur". The Hurian elements and the Indo-Europeans arrived in the Levant during MB II, but the dominance of the Canaanite language and the continuation of the Western-Semitic culture are evidence of the fact that the composition of the local population did not change as much. During this period, the assimilation of the Amorite into the local culture can also be found in Mesopotamia (Issar, 1990), where the newcomers, who were named Amorites by the Sumerians, displaced the Sumerian language but very soon adopted the local Sumerian and Acadian cultures, which replaced the original Amorite culture. It is still being debated whether the newcomers in Canaan were also Amorites who had come from the margins of the Syrian desert.

From the Egyptian *Execration Texts*, which referred to rulers in opposition to the Egyptian dynasties, we learn that the names of all the rulers of the Canaanite cities were Western-Semitic. The peoples in these texts are not defined as Amorite but as "Amu", which certainly denotes an Asiatic origin of the names (Dever, 1976). During MB I, the relationship between Canaan and Egypt strengthened. There were good trade connections and the Egyptian cultural and political influence spread over all the region. Towards the end of MB I, the strong relationship between Canaan and Egypt deteriorated (Aharoni, 1978a,b).

The type of fortification (*glacis* and *terre pisee*) of the MB II and MB III symbolized the new era of walled cities. The emergence of fortification reached its climax precisely during the Fifteenth or the Hyksos Dynasty in Egypt. Dever (1987) suggested that the Canaanite city-states needed a heavy defense system because of the growing strength of the Hyksos rulers in Egypt. The fortifications were intended to defend the city-states against Egyptian expansion toward the north. It is less reasonable to think that the fortifications were built against the threat of local city-states. The connection with Egypt is clear, even though the Canaanite rulers kept their independent status until the end of the MB III (Neeman, 1982).

After the end of the second intermediate period and the collapse of the Hyksos rule (*c.* 3.55 ka BP), all sites of the MB III in Canaan show signs of destruction (Dever, 1987). The massive fortifications could not stop the Egyptians troops, who pursued the Hyksos on their way back to their homeland. It is reasonable to think that the rulers of Canaan and Syria, who were allies of the Hyksos, opposed the Pharaohs of the Seventeenth and the Eighteenth Dynasty and cooperated with the Hyksos (Dever, 1985b; Neeman, 1982; Weinstein, 1981). The Egyptian invasions led to disaster and disorder. There again was evidence indicating the collapse of a very flourishing civilization at the end of MB III. The devastation of the country and the abandonment of most of its sites lasted for about 100 years. Only towards the end of the Late Bronze Age was there a full recovery of the urban centers.

From the proxy-data time series, one can conclude that the MB was a relatively warm period. Yet, according to the same data it was not as warm as EB IV but resembled a continuation of a warm phase. Deposition of sands and sandy silts in the bed of Nahal Lachish during MB I and MB II may also be interpreted as "rapidly fluctuating rainfall patterns interspersed with drought leading to soil stripping from the hill slopes" (Rosen, 1986, pp. 56–57).

Neumann and Sigrit (1978) undertook a survey of references to barley harvest dates in the clay tablets of ancient Babylon. They concluded that the period 3800–3650 BP was warmer than at present, causing the harvest to start 10–20 days earlier than currently feasible.

The renewal of the fortified cities after the disaster of EB IV, despite the tendency of the environmental data to indicate moderate climatic conditions, raises the question of whether the new development was entirely a consequence of socio-economic conditions, or whether it was partly a result of a slight improvement in climate. It has to be noted that the flourishing settlements were those in the center and the northern parts of the country. The Negev remained almost empty, with a small semi-nomadic population, suggesting a rather warm dry climate with some showers from time to time.

The archaeological remains of sophisticated rainwater management and water storage systems in the MB cities indicate a technological knowledge that helped the people to survive. The technology of excavating wells in the rock was probably first known in the MB (Miller, 1980). The techniques of cutting through rock to tap aquifers were known in Egypt and Mycenae during the late fourth millennium BP and were put to use in the Levant, which was influenced by those countries. According to Dever (1987, p. 159) "the very location of Middle Bronze settlements themselves is

ample evidence. They are situated in well-watered regions along the coast, in the river valleys, and in the hill country – always within range of extensive arable lands". It seems that the MB settlements were located in regions with 400 mm yearly rainfall or more. Therefore, the emergence of cities in the MB was more a consequence of environmental adaptation than of favorable climatic conditions.

I believe that the evolution of fortified urban centers, amid agricultural areas, was particularly linked with the acquisition of the skills required to excavate deep wells and tunnels. This enabled the dwellers of the cities to maintain their water supply during periods of war and siege; during periods of peace, it assured a supply of water for the agricultural terraces surrounding the cities, even during years of low precipitation (Issar, 1990). That such dry years did occur can be detected by detailed analysis of the ^{18}O and ^{13}C in the stalagmites of Soreq Cave, in the vicinity of Jerusalem (Bar-Matthews *et al.,* 1998a,b).

1.4.5 The Late Bronze Age and Iron Age, 3.5 ka to 2.6 ka BP

The different proxy-data time series for this period provide rather inconstant evidence with regard to the climate. While the period started with a decline in the Mediterranean Sea level and a rise in the number of planktonic foraminifers, parallel to a rise in the level of the Dead Sea, it continued with a reverse trend, which is inconsistent with most of the environmental and isotope composition data of the lakes and caves deposits. It is, therefore, suggested that this was a period characterized by rather frequent changes from cooler and more humid weather to warmer and drier periods, and vice versa. However, because of the low resolution of most of the time series, except those of Soreq Cave and Lake Van, the imprints of the short-term changes mix and obliterate the details. Consequently, it was decided to reduce the influence of the imprecise data on the interpretation of the impact of climate changes on the hydrology and socio-economic systems in the Levant for this period by relying mainly on the time series of Soreq Cave and Lake Van, and especially on the interpretation of the Soreq Cave section into precipitation values (Bar-Matthews *et al.,* 1998a,b; Fig. 1.3).

Starting *c.* 3.5 ka BP, one can conclude from the proxy-data that the period was rather dry: the Dead Sea was relatively low and the Mediterranean Sea was high. At *c.* 3.4 ka BP, the trend changed towards an increase of precipitation, higher Dead Sea level and lowering of Mediterranean Sea level, which continued until *c.* 3.0 ka BP. Kay and Johnson (1981) found the same trend for the Tigris–Euphrates peak stream flow. Their conclusions are based on a variety of paleo-environmental data, such as Persian Gulf sediments, pollen assemblages from several places in the Near East, Lake Van sediment rates, lake levels, ^{18}O composition and barley harvest dates. According to their calculations, a sharp increase of peak stream flow occurred at about 3.45 ka BP, with a maximum peak at about 3.35 to 3.25 ka BP (1350–1250 BC). Degens *et al.* (1984) detected a rise in the level of Lake Van around 3 ka BP. While the dry period covered mainly the Late Bronze Age, the more favorable climatic conditions covered the period of transition from the Late Bronze to the Iron Age as well as the Iron Age itself.

From the historical point of view, Canaan became part of the New Egyptian Kingdom during the EB, following the victory of the Egyptian kings of the Eighteenth Dynasty over the foreign Hyksos rulers. The zenith of the Egyptian rule in Canaan occurred in the fifteenth century BC (3.5 ka to 3.4 ka BP). During that period, the Egyptians expanded towards Syria, the Lebanon coast and Canaan. At the same time, the Mittani kingdom in northern Syria, and the kingdom of the Hurians, expanded and strengthened along the Upper Euphrates and became the biggest threat to the Egyptian Army (Gonen, 1992). During the fifteenth century BC, Egypt was forced to repress some revolts by the kings of Canaan and Syria, which were supported by the Mittani kingdom. Several expeditions by the Egyptian army to Syria and Canaan were needed to keep its dominion over the area.

During the thirteenth century BC, the Egyptian kings were forced to fight, not only against their traditional enemies in the west, the Libyans, but also against new elements in the area: the "tribes of the sea" and the Israelites (Gonen, 1992). The origin of the "tribes of the sea" is not very clear. But it seems that they came from areas such as the Aegean Sea, the Mediterranean islands and the western and southern coast of Anatolia. At the end of the thirteenth century BC, the Israelites first appeared as a political issue in the area. This period also saw the fall of the great kingdoms, such as that of the Hittites and Ugarit and Alalach, important city-states in Syria. Assyria and Babylon became weakened and Egypt broke down into political segments without a central regime. Egypt lost its influence in Canaan during the second half of the twelfth century BC (Aharoni, 1978a,b). The fall of the great kingdoms was followed by cultural and ethnic changes, brought to the area by new elements like the "tribes of the Sea" (the Philistines) and by the Israelites. The establishment of small states characterized the end of the Late Bronze Age and the beginning of the Iron Age.

The impact of a more humid period may have been why archaeologists have found evidence suggesting a trend towards permanent settlements *c.* 3.4 BP. These semi-nomadic tribes, which may have included the Apiru (Hebrews?), began to settle down along the desert margins, namely the areas where the Canaanite settlement was not dense and strong, especially the mountainous areas, the Negev and central and southern Trans-Jordan. The settlements were also far from the Egyptian centers in Canaan.

The settlement in the mountains at the end of the thirteenth century BC seems to have been partly the result of adapting a new Canaanite invention – the cistern which consisted of an underground storage chamber, excavated into impervious layers or caulked to prevent loss by seepage, and naturally roofed to prevent evaporation. The new water system enabled the establishment of small independent settlements that were not limited by the number and location of the springs. The new Israelite settlements in the mountainous areas brought the biggest settlement revolution in the history of the land of Israel. The Negev mountains also started to flourish (Cohen, 1989). The settlements of Iron Age I (3.2 ka to 3. ka BP) were unwalled farms compared with the fortified cities of the next period, the Iron Ages II and III (3 ka to 2.6 ka BP). It was the first time that settlements were established far from springs or any natural water resources and water supply was ensured by the collection of rainwater from a catchment area into cisterns (Aharoni, 1978a,b).

As already discussed, the Soreq Cave and Lake Van findings show a warming and drying phase before and some time after 3 ka BP. Neumann and Parpola (1987), reviewing textual and non-textual environmental evidence, concluded that the period from 3.2 ka to 2.9 ka BP was warm and dry. The widespread migration of "Sea People", who swept the region at *c.* 3.2 ka BP, may be linked to this dry phase prior to 3 ka BP. This wave of tribes come, most probably, from the Aegean basin. Also, Weiss (1982) suggested that deterioration in climate was the main reason for the immigration of tribes and the collapse of old civilizations. He styled the years from *c.* 1200 to 825 BC as the Ancient Dark Ages. Civilization regressed almost everywhere throughout the eastern Mediterranean and Near East from Late Bronze Age levels. If this is not generally true for the Levant, where small states were able to prosper occasionally, it is only because there was a singular absence of any major power to interfere in their affairs (Weiss, 1982, p. 182).

At about 3.2 ka BP (1227 BC), Libyan tribes from Cyrenaica invaded the Nile delta but were defeated by the Egyptians (Carpenter, 1966). The Hittite kingdom, which had gained its strength during the Bronze Age, collapsed and disappeared (Carpenter, 1966; Weiss, 1982). The destruction of this kingdom was connected to the great movement of tribes out of the forest-lands of Europe or the steppe lands of Asia. But Carpenter (1966, p. 46) points that "even if they [the tribes] had combined their forces, these miscellaneous coastal tribes would not have had strength to destroy the great Hittite empire".

Big city-states along the Syrian coast, such as Ugarit and Alalach, were suddenly destroyed and never reoccupied. The island of Cyprus was ravaged around the year 1200 BC. Assyria, too, underwent several revolts, which weakened the empire. The very important cities in Canaan, such as Hazor, Lachish, Beit Shean, Megiddo, Afek, Beit Shemesh, Gezer, Tel Beit Mirsim and others, were destroyed (Mazar, 1990).

The transition from the Late Bronze Age to Iron Age I in the Levant was characterized by a new ethnic-political structure, which brought several ethnic states into the area: the Philistines and other Sea Peoples, the Israelites, the Phoenician–Canaanites, the Edomites, the Moabites and the Ammonites. The beginning of the twelfth century BC, therefore, represents the fall of the big kingdoms and the establishment of the small national states.

Canaanite cities continued to exist during the twelfth century BC. The concentration of settlements was in the northern valleys (Megiddo, Beit Shean). It may be that these Canaanite settlements existed under the rule of the Philistines. In the north, a new Phoenician culture was established, based on the roots of the Canaanite culture (Mazar, 1990).

The Israelites first occupied the unpopulated areas in the mountains, leaving alone the Canaanite cities in the valleys. It seems that the Israelites avoided military contact with the local population, at least in the first stage of settlement. Therefore, they settled in the marginal areas, while the Canaanites lived in the valleys and the other new element in the area, the Sea People, occupied the coastal plain. Areas such as Trans-Jordan, Gilead and Galilee, Efrayim and Menashe mountains were settled rapidly and very intensively (Aharoni, 1978a,b, 1986).

When one goes to the borders of the desert to look for the impact of climate changes, one finds that new settlements occurred at c. 3 ka BP in the Negev mountains that were a complex of fortresses and agricultural farms (Aharoni, 1979a; Cohen, 1989). There are still basic disagreements concerning the exact date of the appearance of the fortifications, the cause of their disappearance and their historical role (for more details see Aharoni, 1967; Cohen, 1989; Finkelstein and Perevolotsky, 1989). Aharoni (1967) connected the building of the fortifications in the Negev mountains to the establishment of King Solomon's kingdom but considered that this building occurred within the whole period between the tenth century BC and the seventh century BC. According to Aharoni, the kings of Israel and Judea built fortresses in order to control the commercial routes to Eilat. Cohen (1979a,b, 1985) claimed that the fortifications were built during the middle of the tenth century BC as a defense system in the southern part of King Solomon's kingdom. The expedition of the Pharaoh Shishak (Sheshonk) from Egypt to Israel (925 BC) ended the settlement in the Negev (Cohen, 1979a,b; Mazar, 1986). Cohen (1979a,b) claimed that, after Shishak's conquest, the Negev mountains remained unpopulated for several centuries, until the Persian period, and the Beer-Sheva valley became the southern border of Judea. Finkelstein and Perevolotsky (1989) suggested that the settlements were built by the people of the desert and were not fortresses at all. Whatever opinion we adopt, Iron Age IIa was a climax of prosperity in the Negev mountain area and the Beer-Sheva valley.

As can be seen from the Soreq Cave and Lake Van data, the dry period was followed, c. 2.8 ka BP, by a more humid period. The impact of the cooler climate may be recognized during the period the archaeologists call Iron Age IIb (2.8 ka to 2.7 BP, 700

to 800 BC), characterized by building and fortification of cities and the notable economic growth of the kingdoms of the Levant. Samaria of the kingdom of Israel and Jerusalem of the kingdom of Judea were relatively big capitals. During this period, the Assyrian empire came to its full strength and expansion. The expeditions of the kings of Assyria to the "west" had already started during the ninth century BC and caused destruction in the northern part of the kingdom of Israel. Towards the end of the eighth century BC, Israel was totally destroyed, and the kingdom of Judea, which was the only area that was not conquered by the Assyrians, was greatly ravaged (701 BC) (Barkay, 1992).

An improvement of climatic conditions towards c. 2.8 ka BP was detected by Neuman (1985), who searched for references to climatic conditions in classical Greek and Roman literature. He came to the conclusion that northern Italy and, therefore, probably the whole Mediterranean area, was affected by a relatively cool and wet climate between 2.8 ka and 2.4 ka BP (the end of the Iron Age and the beginning of the Persian period). Based on the massive building activity in Israel, it might be reasonable to think that the improvement in climate began some time after 3 ka BP. Schaeffer's findings at Ugarit, Enkomi and Cyprus (1968, cited in Neuman, 1985) also favor this conclusion.

The last period of the Iron Age is Iron Age III (700–586 BC). Areas in Israel and Philistine had already been under direct rule of the Assyrians for about 100 years. The Assyrian kings had largely destroyed Judea at the beginning of the seventh century BC, but later on, from time to time during the seventh century BC, Judea flourished under the rule of some of the Judean kings. From the end of the seventh century BC until the beginning of the sixth century BC, many Judean cities were destroyed by either the Egyptians or the Assyrians. At the beginning of the sixth century BC, Israel and Judea were sandwiched between two great powers: Egypt and Babylon. By the end of the period, Israel and Judea had been conquered by Nebuchadnezzar, the King of Babylon. However, even though the Babylonians strengthened the fortification of the Negev, this area was conquered by Edom by the end of the period.

A short dry period, at around 2.6 ka BP, was followed by a rather humid period, which lasted until c. 2.2 ka BP. During this time, the Levant was ruled by the Babylonians (586–538 BC), the Persians (539–332 BC) and the Hellenistic kingdoms of Alexander's inheritors, the Seleucids in Syria and the Ptolemeans in Egypt. This humid period may explain why the area still flourished, notwithstanding the political crisis in the Levant followed by the conquest and subordination by Babylon. Moreover, it may explain the flourishing of Persia, which was mostly desert (as now) and partly semi-arid and depended mainly on rainstorms coming in from the Mediterranean.

In Judea, the political and cultural crisis in 586 BC, caused by the destruction of Judea and Jerusalem by Nebuchadnezzar, was not as severe as first thought (Barkay, 1992). In the north of the country, the coastal plain and the Negev, there was a continuation of settlement, even after the Babylonian conquest. Areas such as the Judean mountains and Hebron were captured after the destruction of Judea by the Edomites, who probably were pushed northward by Arab tribes (the "fathers" of the Nabateans). The Philistines occupied the coastal plain, although parts of Judea were still settled by the people of the tribe of Benjamin. During almost all the Babylonian period, the Philistine land (south of Jaffa) was a semi-autonomous area, with several city-states that were vassals of the king of Babylon (Liver, 1986). Barkay (1992) claims that the material culture of the Babylonian period continued to be that of the Iron Age.

In the marginal areas in the east, there were three national states: Moab, Edom and Ammon. These states suffered from attacks of nomadic tribes, which had been weakening them since the sixth century BC. After the death of Nebuchadnezzar (562 BC) and until the conquest of Babylon by Cyrus the King of Persia (539 BC), the Babylonian kingdom and its provinces suffered from instability.

In its early days, the Persian empire included the Iranian plateau and the entire Babylonian kingdom. Later on, the Persians conquered northern India, Asia Minor and Egypt, and the Persian empire became the greatest empire in the ancient world.

During the Persian period (539–332 BC), Israelites who were allowed to return from exile settled in the northern part of Judea. There was also a settlement on the border of the Negev, and Gaza became the central city in the area. The city of Beer-Sheva was renewed and a fortress was built. The range of the settlement was, however, not as large as before (Cohen, 1979a). Some other sites in the Negev from the Persian period are Ezion Gever (Tel El-Halifa), Tel Arad, Retema and Kadesh Barnea (Cohen, 1986; Meshel, 1977). Meshel (1977) claimed that the location of the settlements was similar to that of the earlier period during the Judean kingdom. The settlements were built along the commercial road that connected the cities of the southern coastal plain with Eilat. Nevertheless, the settlements of the Persian period were not sedentary settlements, such as the ones during the previous Israelite and subsequent Nabatean periods. The principal function of the settlements in the Negev at that time was to store grain and water. Stager (1971) described the storage pits found at Tel el-Hasi and some other sites located in the marginal areas of the Negev (at present 300–400 mm mean annual rainfall). Stager (1971) cited Petrie (1928), who concluded that the Tel Jemmeh site was a supply post for the Persian army attacking Egypt via Sinai during the Persian period (fifth to fourth centuries BC). Moreover, the building of huge underground storage bins in this marginal area was interpreted by Stager (1971, p. 89) as a necessity because of the constant threat of drought: "To cope with the 'lean' years of harvest, the communities needed facilities for storing a surplus during the seasons of plenty".

The Edomites settled south of the new Israelite settlements, after they were pushed northward by Arabian tribes (Liver, 1986). Parts of Trans-Jordan (the Gilaad) were also settled. The Phoenician cities flourished during the Persian period (Aharoni, 1979a).

During the fourth century BC, the Persian empire declined. Riots spread all over the kingdom and forced the Persians to fight against Egypt, Phoenicia, Cyprus and even Israel.

Since the conquests of Alexander the Great (332 BC), the Greeks had become a dominant element in the East. The new Hellenistic period introduced Hellenic culture into the Levant. Even though the Macedonians had had great success in Egypt and Asia, they were under the threat of other nations such as those from Europe and the Middle-Asian nations in the East (Stern, 1986). It is interesting to note that the tendency of the Greeks to emigrate to the East during the Hellenistic period was the result of severe social and economical crises in Greece.

1.4.6 The Roman–Byzantine period (2.3/2.2 ka to 1.3 ka BP), including the Roman–Byzantine transition period (1.7 ka to 1.6 ka BP)

From about 2.3 ka BP to about 1.3 ka BP, there is quite a good correlation between the different proxy-data time series in the Levant. It started with a phase of cooling and humidity which continued up to c. 1.7 ka BP, during which time the level of the Dead Sea rose to more than 30 m above the present level, there was a profusion in planktonic foraminifers in the Eastern Mediterranean Sea, and a lighter composition ratio of environmental isotopes were laid down in cave and lake deposits. Degens *et al.* (1984) observed a rise in the level of Lake Van during this period.

Then, at c. 1.7 ka BP, there was a reversal of these trends, indicating a warmer and drier period, which continued up to 1.6 ka BP and was followed by a colder and more humid period, which lasted for two centuries. Degens *et al.* (1984) observed a rise in the level of Lake Van around 1.5 ka BP. Then, at c. 1.4 ka BP, a severe change occurred to a period of warming and desiccation. As this period was that of the hegemony of Rome and later Byzantium over the Levant, it is suggested that it should be called the Roman–Byzantine period.

During most of this period, there was a profusion of settlements in the semi-arid plain of Beer-Sheva (Issar *et al.*, 1992). Liphschitz *et al.* (1981) studied the assemblage of trees from the Roman siege ramp in Massada (c. 70 AD) and found that the ratio of tamarisk to acacia was higher then than at the present. This, in their opinion, indicates that wetter conditions existed during the first century AD, as the tamarisk needs a higher groundwater table than acacia. Yakir *et al.* (1994) confirmed this conclusion by comparing the isotopic ratios ($\delta^{13}C/^{12}C$ and $\delta^{18}O/^{16}O$) of the cellulose of the tamarisk trees from the same siege rampart with isotopic ratios of the cellulose in present day tamarisk trees growing in the Massada region and central Israel. They found that the ancient tamarisk cellulose was depleted in both ^{13}C and ^{18}O compared with cellulose from trees growing in the same region today, and the ancient trees were similar in their composition to trees from the northern part of Israel. While lighter ratios of oxygen isotopes are an indication of a lighter composition of the rain and less-extreme evaporation processes, the lighter composition of carbon isotopes indicates better conditions of water supply, which meant that the trees did not need to close their pores to reduce transpiration in times of insufficient supply of water. The isotopic data suggest that relative atmospheric humidity in the Roman period was about 15% higher than today (Issar and Yakir, 1997).

Samples of ancient tamarisk wood, which were obtained from the study carried out in the caves of Mount Sodom (Frumkin *et al.*, 1991) and dated to the Roman period, were similar to the wood from Massada, while the earlier samples, dated *c.* 4 ka BP, showed ^{13}C composition similar to that of the trees around Massada of present day, indicating a dry climate at that time.

Palynological data of Baruch (Stiller *et al.*, 1983–84; Baruch, 1986), taken from a core in the Sea of Galilee, shows a conspicuous increase of olive pollen and a decrease in oak and pistachio during the Roman period. Baruch concluded that the economy flourished during the Roman period and, consequently, farmers reduced the natural forest to replace it with olive groves. According to Baruch, the decline in olive pollen at *c.* 500 AD was caused by an economic crisis in which inflation increased, causing the export of oil to be too expensive. Yet, as already discussed, Baruch's curve also shows olive increase and oak and pistachio decrease during the EB period, together with a reverse of conditions at the end of the period, earlier shown to be in accord with proxy-data and historical evidence for climate changes. Furthermore, depletion in ^{18}O and ^{13}C isotopes in lacustrine carbonates of the Sea of Galilee indicates a colder climate during the Roman period.

Klein (1982) mapped the ancient Dead Sea shorelines all along its western bank and dated them according to archaeological sites. According to her findings, the sea rose to about 70 m above the present level at *c.* 2 ka BP. If these levels are entered into the hydrological computer model of the Water Planning for Israel Company, which simulates the hydrological balance of the Dead Sea drainage basin, an increase of about 40% in precipitation was calculated. Although this simulation did not take into consideration the reduction of evaporation that would occur in the cooler climate, it is still obvious that there was an increase in the rate of precipitation. Bar-Matthews *et al.* (1998a,b) calculated the level of precipitation over the Judean mountains based on the speleothemes research in Soreq Cave. They concluded that, *c.* 2.2 ka BP, average annual precipitation reached 530 mm, compared with 500 mm today, namely about 6% more. Klein's shoreline observations

were later partly confirmed by Frumkin *et al.* (1991), who examined caves on Mount Sodom. They also found a higher shoreline, not as much as 70 m but to about 30 m above the present level.

Nir and Eldar (1987) and Nir (1997) investigated ancient water levels in wells along the Israeli Mediterranean coastline region. In the paper of 1987, the research was looking for evidence for tectonic movements, while in the paper of 1997 the fluctuations were believed to "reflect historic eustatic sea-level changes and the rate at which the end of the post glacial transgression advanced". I suggest that they were actually describing the impact of paleo-water tables, which would be influenced by both rates of recharge and sea-level changes, when the former may have even masked the latter. This masking may have happened in periods when precipitation, and, therefore, groundwater recharge, was at its maximum or minimum, similar to what contemporaneous water tables show during good or bad years.

Sanlaville (1989) observed the retreat of the Persian Gulf sea level *c.* 2 ka BP, namely during the Roman cold period (which he terms Hellenistic). Bernier *et al.* (1995) observed a similar general pattern of shoreline movement in a lagoon off the Persian Gulf in a part of the United Arab Emirates. The lowest level was reached *c.* 2 ka BP.

Of special interest, from the point of view of tracing the impact of climate change on the hydrological cycle, is the history of the Nabatean people. They settled in the area bordering the Levant and the Desert of Arabia some time during the fifth century BC and developed a sophisticated system of rain-fed agriculture, until they disappeared during the seventh century AD. The origin of the Nabatean and their routes through the desert until they reached Trans-Jordan and Israel are unclear. Their recorded history starts during the late Persian period and the early Hellenistic period (fifth century BC till 100 BC), when an organization of nomadic tribes, who had come from the desert, established a kind of "nation" in the Edom and Negev highlands (Negev, 1965). Negev (1977) cites Daidorus and Hieronimus, who described the early Nabateans as nomads who lived in areas characterized by "lack of water" and who raised camels and sheep. They also controlled the trade routes and specialized in transferring frankincense and myrrh, which were brought from southern Arabia to the Mediterranean ports. Another Nabatean occupation was the export of bitumen from the Dead Sea to Egypt.

Their knowledge in rainwater harvesting seems to have been very advanced. They knew how to excavate underground reservoirs in impermeable rocks; these reservoirs could reach 30 m depth. After the reservoirs were filled by rainwater, the Nabateans covered them and kept them secret from their enemies. In this way, they were able to survive in dry areas such as the central Negev. They later developed a sophisticated methodology of water harvesting for conveying surface flow from hillside slopes into terraced riverbeds (Evenari *et al.,* 1971).

At about 100 BC, they came into conflict with the Hasmonean–Jewish kingdom, and the Nabatean settlement in the Negev disappeared. This deprived the Nabateans of access to the Mediterranean ports and forced them to shift east and northwards and conquer northern Syria and Damascus. The first century BC was a transition period in the history of the Nabateans when they became a semi-sedentary society. At the end of the first century BC, and the beginning of the first century AD, the Nabatean settlement reached its climax. At this point, the renewal of their settlements in the Negev desert started, cities were established (Halutza, Avdat, Nessana, Mamshit, Shabeita and Rehovot), as well as a network of fortresses along the way from Petra, their capital in Trans-Jordan, to Gaza.

On one hand, the growth of the demand of the Roman market for spices, perfume and wood from India strengthened the economy of the Nabateans yet on the other hand, the Romans understood the power residing in control over the trade routes and tried to take that control. The peak of Nabatean prosperity was at the beginning of the first century AD, during the reign of Hartaat IV (9 BC to 40 AD). Most of the archaeological remains of buildings, arts and Nabatean inscriptions are from that period, a time of peace between the Nabateans and the kingdom of Judea, during which the Nabateans did well on the trade from Arabia. After the death of Hartaat IV, the Nabatean kingdom started to decline. During the reign of Malho II (40–70 AD), it became a vassal kingdom of Rome. Nevertheless, during the reign of the last king of the Nabateans (Ravaal II, 70–106 AD), their agriculture continued to prosper. In 106 AD, after the death of Ravaal II, the Romans took over the Nabatean kingdom and annexed it in their "Provincia Arabia". This province included Trans-Jordan and the former Nabatean kingdom in the Negev.

Still, the Nabatean settlement in the Negev began to recover. The main road connecting the Red Sea with the Mediterranean ports regained its importance (106 AD till the end of the third century AD). By the end of the third century AD, there was a full assimilation of the Nabatean culture with the Roman culture (Negev, 1965, 1977, 1979). During this period, the Negev became part of the Roman *limes* system, and border guards, the *Laminate*, settled all over the border region. From the Nessana papyri, which were discovered by the archaeological expedition headed by Colt in 1936 (Lewis, 1948), it seems that Nassana belonged to this defense system. The military stations had been constructed by the end of the third century AD. The *Limitanei* were not active soldiers. They lived with their families near the frontier, and the economic basis of their life was agriculture. The civilian areas included cities and tribal villages. The process of urbanization

took place during the Byzantine period. All over the Negev, there is plenty of evidence of intensive agricultural activity near the urban centers. The terraces were mainly vineyards, irrigated by the sophisticated water-harvesting systems. But even though they cultivated the land using the very advanced systems, they could not supply all their needs and had to import wheat from Egypt and the Gaza region (Avi Yona, 1934, 1977).

A very interesting description of the physical environment at the end of Byzantine period can be found in a letter from the sophist philosopher Procopius (*c*. 450–526 AD) to his friend Jerome in Egypt: "There will be a day when you will see Elusa again and you will weep at the sand being shifted by the wind stripping the vines naked to their roots." (Cited in: Mayerson, 1983, pp. 251–253.) From the letter, one can conclude that, in the first half of the sixth century AD, the settlers in the northern Negev, where at present the average annual precipitation is about 80 mm, still raised vineyards, even though the physical conditions had started to deteriorate.

1.4.7 The Moslem–Arab period, *c*. 1.3 ka to 1.0 ka BP

The deterioration of the climate all over the Middle East sometime around 13 ka BP can be deducted from most of the proxy-data time series, especially from the environmental isotope composition of the deposits of the Sea of Galilee, the caves of Galilee and Soreq Cave. Bar-Matthews *et al.* (1998a,b) calculated, based on the isotopic composition of the stalagmites in Soreq Cave, that the average annual precipitation over the Judean mountains had begun to decrease some time after 1.4 ka BP (after reaching the high level of 530 mm, compared with 500 mm at present) and reached a low of 450 mm some time around 1 ka BP.

Raban (1991) and Raban and Galili (1985) indicated that the level of the Mediterranean Sea rose towards the end of the Byzantine and during the Arab period. As already mentioned, Nir and Eldar (1987) observed that the ancient water table levels reached a maximum during the Roman–Byzantine era and receded during the Moslem period. Sanlaville (1989) showed that the Persian Gulf was at a high level *c*. 1.3 ka BP and Degens *et al.* (1984) observed a recession of the level of Lake Van *c*. 1 ka BP.

In my PhD thesis, I examined the young sand dunes covering the western part of the coastal plain of Israel, which were of post Byzantine age, and suggested that the invasion by the sand dunes occurred because of destruction of agriculture by the Arab conqueror coming from the desert, lacking knowledge of agricultural management. Following the interpretation of the data from the Sea of Galilee core, Galilee caves, Mediterranean and Dead Sea levels etc., it seemed more likely that the invasion of the sand was a result of its higher rate of supply from bigger Nile floods, as can be seen from its higher levels during this period (Nicholson, 1980), and the erosion of the Nile delta by higher sea levels. Such a process of invasion by dunes also occurred during the warm period at the end of the EB (Issar and Tsoar, 1987; Issar *et al.*, 1989; Issar, 1995a,b). Indeed, according to historical evidence, the Umayyad caliphs kept the Byzantine settlements and preserved their Christian administration.

Tsoar and Goodfriend (1994) suggested that the activation of the dunes, which they observed by dating hearths in the dunes, also occurred from 600 to 900 AD and from 1700 to 1950 AD was a function of overgrazing and trampling. The precipitation curve from the Soreq Cave (Fig. 1.3) shows that, as in the Chalcolithic period, these periods were characterized by a shift from a peak of precipitation to desiccation. I would suggest that here also the primary factor for the activation of the dunes was the aridization phase, which immediately followed the peak of humidity, namely the Moslem–Arab dry period, which followed the Byzantine humid period, and the Ottoman dry period, which followed the Little Ice Age humid period.

The Arab conquest did not cause the immediate disappearance of settlements. According to the papyri of Nassana, the settlements existed unchanged at least to the very end of the seventh century AD and were abandoned in the first half of the eighth century AD. Only some pieces of pottery and nine Umayyad coins before 750 AD have been found (Lewis, 1948). Also none of the Greek papyri are dated after 700 AD. According to an archaeological survey of the Negev described by Cohen (1985), only six Arab sites were discovered from the early Moslem period of the Umayyad rule compared with 44 Byzantine sites. He doubts whether these sites continued to exist during the rule of the Abbasid House. The geographer Huntington (1911), who traveled all over the Middle East, was the first to suggest that change of climate caused the desertification of this region during the Moslem-Arab period. Also, Kedar (1985 p. 15), in his study of the Arab conquest and agriculture in the seventh century AD, suggested that regional climate may have deteriorated in the seventh century; that this deterioration was one of the causes for nomad migration into settled lands; and the same deterioration, along with nomad incursions, might have affected the vegetation of the conquered countries.

Most of the contemporaneous archaeologists, historians and geographers prefer, however, anthropogenic explanations for this colossal desertion (Avi-Yona, 1934; Evenari *et al.*, 1971; Sharon, 1976; Tzafrir, 1984). Yet, my survey showed that most, if not all, of the settlements that were totally deserted were located to the south and east of the 200 mm/year precipitation line (including the city of Beer-Sheva, which flourished during Byzantine times). In conjunction with the proxy-data time series already discussed, this leaves little doubt that the basic cause was indeed that of climate change.

Carpenter (1966) explains the desertion of many monasteries, scattered all over Anatolia and Greece during the seventh century AD, as a result of an economic crisis caused by a climate change giving rise to frequent droughts.

Lev-Yadun *et al.* (1987), investigating annual tree rings as an index of intensity of climate changes in the Levant, found that the most suitable trees in this region for dendrochronology are Red Juniper, Persian Pistacio and Cedar of Lebanon. The rings of the Jerusalem Pine, for example, show that the range of changes in rainfall is about 400 to 900 mm per year. The changes are not linear and there is no growth at all if rainfall is less than 400 mm per year. Research into the rings of the Cedar of Lebanon and Red Juniper trees show that there have been only slight changes of weather in the last 2000 years, some of which were regional and some only local. From *c.* 420 to 480 AD, tree rings were very narrow, pointing to drought.

1.4.8 The Crusader period and Little Ice Age, 1.0 ka to 0.4 ka BP

The isotopic data from the deposits of the Sea of Galilee, the caves of Galilee and Soreq Cave show a trend of depletion *c.* 1 ka BP. Also, the level of the Dead Sea rose. More accurate dating of the speleothemes of Soreq Cave and interpretation of the findings with regard to levels of precipitation indicate that the climate became colder and more humid immediately after *c.* 1000 AD onwards, reaching, with some variations, a peak *c.* 1600 AD, when the average precipitation on the Judean mountains was 570 mm. (Bar-Mathews *et al.*, 1998a,b). At about this time, the level of the Mediterranean Sea showed a marked retreat and the level of Lake Van rose, at *c.* 0.5 ka BP (Degens *et al.*, 1984).

1.4.9 The Moslem–Ottoman period, 0.4 ka to 0.1 ka BP

Through better resolution of the time series of Soreq Cave, it can be seen that around 0.4 ka BP there was again a short but severely warm and dry period, lasting about a century, which reduced precipitation to an annual average of 450 mm. Thereafter, the climate became cooler and more humid, exceeding the current average rainfall.

Liphschitz and Waisel (1974) explained the disappearance of date palms from the region of Jericho and Ein Gedi from the fifteenth century AD as a result of human activity. As the date palm is a tree that needs an ample supply of water, one can question whether the basic reason for the date palm's disappearance was the reduced amount of water supplied by the springs in these two places.

A more precise picture of the changes during this period can be reached from tree ring investigations. Liphschitz *et al.* (1979a)

Table 1.2. *Radial growth of* Pinus nigra *in the period 1670–1950 AD*

Trees with common covariance		
	Wide rings	Narrow rings
Group 1	1670–1710	1720–1740
	1800–1820	1830–1850
Group 2	1740–1760	1780–1800
	1920–1950	1880–1910

investigated the dendrochronology of *Pinus nigra* in southern Anatolia (Turkey). This species is found in the Mediterranean region at altitudes between 300 and 1800 m, where the annual amount of precipitation is above 1000 mm. The tree produces distinct annual growth rings and attains ages of several hundreds of years. Samples from the Taurus mountains, 50 km north of Karsanti, at an elevation of 1775 m, where annual rainfall varies between 735 and 1400 mm show that radial growth in *P. nigra* depends on temperature conditions in certain months, more than on any other climatic factor (Table 1.2).

Dendrochronological investigations on *Juniperus polycarpos* growing in west and central Iran revealed that the radial growth in this species depends in the more arid regions mainly on the amount of precipitation (Liphschitz *et al.,* 1979b). When the amount of rain is sufficient (i.e., above 450 mm), the prevailing summer temperature seems to become the limiting factor. Wide rings (favorable conditions) were found in 1685–95 and 1790–1800. Narrow rings (less favorable conditions) were found in 1725–35 and 1855–65.

Serre-Bachet and Guiot (1987) investigated summer temperature changes using tree ring analysis in the Mediterranean area during the last 800 years. Their conclusions are as follows: 1150–1250 intermediate, 1250–1340 warm, 1340–1400 cold, 1400–55 intermediate, 1455–1550 warm, 1550–1605 cold, 1605–25 warm, 1625–50 cold, 1650–85 warm, 1685–1705 cold, 1705–40 warm, 1740–60 cold, 1760–1810 warm–intermediate, 1810–40 cold, 1840–present, many fluctuations.

1.4.10 The Industrial period, 0.1 ka BP to the present

Striem (1985) investigated the quantitative and qualitative aspects of recent climatic fluctuations in Israel. This study was based on systematic meteorological observations, which have been made in Jerusalem since about 1850, and these constitute the longest written record of climatic parameters in the Middle East. He found that winter (December to March) temperatures rose by more than 1 °C, while the summer (June to September) temperature remained steady. The mean monthly barometric pressure rose by about 1 mb. The annual rainfall decreased from about 600 mm in the second half of the nineteenth century to about 500 mm in the first half of the twentieth century; towards 1985, there was a trend of increased rainfall. The pressure maximum used to occur before November (44% of years in the period 1861–85), but from 1946 to 1970, it was only in 17% of years that the pressure maximum occurred before November. During these years, its occurrence in November became more frequent (an increase from 24 to 46%). The retardation in the occurrence of the barometric pressure maximum is characteristic of drought years.

A series of warm years occurred between 1920 and 1963. After 1930, the level of the Dead Sea dropped by about 2 m from a level of around 390 m BSL, which had been maintained since 1893, to a level around 392 m BSL (Ashbel, 1938). The Dead Sea remained at the latter level, albeit with fluctuations, from 1940 until 1959, after which its level has continued to decline.

Van Zeist and Bakker-Heeres (1979) investigated the economic and ecological aspects of plant husbandry of Tel Aswad, situated some 30 km east-southeast of Damascus. Mean annual rainfall is below 200 mm at present. The subsoil in the Aswad area consists of lacustrine sediments, deposited in a lake, which, up to Late Pleistocene times, must have covered a large part of the Damascus basin. The natural vegetation is a steppe. It is likely that the lake and the marshes extended up to the site of Aswad in early Neolithic times, as it did around 1850 AD. In 1852 AD, the lake also extended to the edge of Tel Aswad. In 1880, the south lake had already receded to some extent and today only part of the north lake of Aateibé remains.

2 Climate changes during the Holocene Epoch in Europe

2.1 CLIMATE

2.1.1 Contemporary climate

The contemporary climate regime of Europe is spatially variable, because of its position between the Arctic and Mediterranean zones on the latitudinal extent, and the Euro-Asian continental mass and Atlantic coast zones on the longitudinal extent. Additional spatial climatic variability results from the topography of Europe, which is characterized by a combination of high mountain chains, low lying countries and a circuitous shoreline. The penetration of the sea into the continent causes the climate to become milder, in terms of temperature. Consequently, the climatic conditions in each region differ according to its position in relation to the more extreme conditions along the borders. Another influence is the warm Gulf Stream, which flows along the western shores of Europe. From the global point of view, the European continent lies within the westerlies wind system. Within this system, barometric pressures influence the storm regime. High-pressure, anticyclone system characterizes the eastern continental region, while the low-pressure cyclone system characterizes the west over the Atlantic.

2.1.2 The Pleistocene–Holocene transition period

The last glacial maximum occurred between 22 ka and 16 ka BP. A massive deglaciation started shortly thereafter, characterized by strong fluctuations and occurring between 15 ka and 8 ka BP. A warm period seems to have taken place from 13.5 ka to 11 ka BP. This is referred to as the "Windermere interstadial" in Britain and Ireland and the "Bölling interstadial" in Scandinavia. A brief cold phase (mini-stadial) occurred within this warm period in the British Isles, around 12 ka BP, equivalent to the Older Dryas in Denmark and Scandinavia. The warm Allerod phase (11.8 ka to 11 ka BP) followed the Younger Dryas (Mangerud, 1987).

Pollen data from Luxembourg, France, Corsica and Spain indicate that a sudden and marked cooling (4–5 °C), as well as a fall in precipitation (400–500 mm), occurred at 11 ka BP in southwestern Europe, with a temperature and precipitation minimum that still persisted at 10.5 ka BP. At this time, temperature was again similar

to that of the Older Dryas, but precipitation was intermediate between that of the Older Dryas and today. Nevertheless, the reduction in precipitation was more marked in winter than in the other seasons. It is likely that the important advance of the glaciers in the Alps was made possible by a medium accumulation of snow in spring and autumn and, more importantly, weakened melting in summer (Pons et al., 1987).

There is evidence of a period of high temperatures c. 13 ka BP (especially during summer, reaching or even rising above those of today), when temperate beetle (Coleoptera) assemblages replaced Arctic ones (Roberts, 1989). As beetles react quicker than plants to temperature changes, the rate of warming c. 13 ka BP at a site in north Wales is calculated to have been an amazingly fast 1 °C per decade (Coope, 1975). Summer temperatures in central Britain c. 13 ka BP, rose suddenly by 7 °C and winter temperatures may have risen by 20 °C at the same time (Coope, 1987). Paleo-temperatures in Britain have been calculated on the basis of fossil beetle assemblages for both summer and winter in the period 14.5 ka to 9.5 ka BP (Atkinson et al., 1987). However, climatic reconstructions based on Coleoptera for the Holocene Epoch have so far proved to be more problematic (Roberts, 1989).

2.2 THE MEDITERRANEAN REGION

2.2.1 The Balkan peninsula

The Mediterranean climate in the Balkan Peninsula is limited to the coastal area and the island region. On the Adriatic coast, the average winter temperature is moderate, except for the northern region where continental cold air is introduced by the Bora wind. In July, the average temperature is 23–25 °C because of maritime influences. In the coastal area, precipitation is heavy. For instance, some places in Yugoslavia receive more than 4000 mm of rain annually. On the western mountain slopes and in low-lying areas, annual precipitation is much less (500–600 mm). In the southern areas, the Mediterranean climate is much more pronounced, with lower precipitation and higher temperatures (for example, Athens never receives more than 400 mm of precipitation annually). The average July temperature in Athens is 27–28 °C, while in January it is 7–8 °C.

The climate on the southern Aegean coast tends toward the continental, with comparatively lower precipitation. In the Aegean islands, the climate is relatively dry, while the climate of the Ionian Sea is more hot and humid.

Paepe and co-workers investigated the effect of changing climate during the Quaternary on landscape and soil profiles in Greece (Paepe, 1984) and claim that there is a periodic cyclicity in the rate of sedimentation. In the Marathon Plain, six Holocene soils were described separated by alluvial deposits. The earliest, Marathon Soil (HS1), most probably developed about 7 ka BP and the last, Kallileios Soil (HS6), is dated 725 ± 5 BC. In the latter soil, geometrical tombs were located. The in-between layers, HS3, HS4 and HS5, together with relevant fluvial gravel deposits, correlated with the three phases of the Helladic period. Soil formations in the fluviatile valley system tally perfectly with peat development in the marine sequence of Marathon. Furthermore, between soil development phases, fluviatile sedimentation rates score the highest values (Fig. 2.2).

Paepe (1984) interpreted the soil layers as indicating warm periods, while pluvio-alluvial periods are cold. Comparing the section of Marathon with the paleo-climate columnar section of the Levant (Fig. 2.2), I would suggest that the cold wet periods would be characterized by rather small sedimentation rates, because the forest expands and there is less soil erosion, while during dry periods, higher rates of soil erosion would cause higher sedimentation rates. This would better explain the correlation of soil formation periods with periods of profusion of settlements.

Of special interest is the observation that, towards the close of the third millennium BC (as determined by pottery finds), heavy destruction is visible over large areas of Greece: "Settlements which were, for their time, rich and powerful, and which had a long history of stability and continuity, literally came tumbling down" (Finley, 1981, p. 13). This period can be correlated with a peak in the deposition rates of pluvial material in the Marathon Plain and, thus, with the warm arid period at the transition from the EB to the MB period in the Levant.

In general, the Marathon Plain sedimentation rates can be correlated with the proxy-data time series of the Levant in the following manner.

During the Neolithic, the Pleistocene forest still dominated the area, protecting the soil cover. Towards the Upper Neolithic, the rates of sedimentation became higher as a result of warming of the climate. The rates fell again towards the Chalcolithic and EB but rose towards the MB. The Iron Age was again characterized

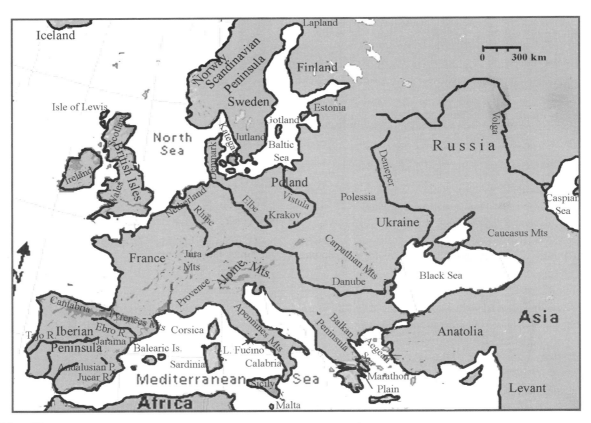

Fig. 2.1. Map of Europe.

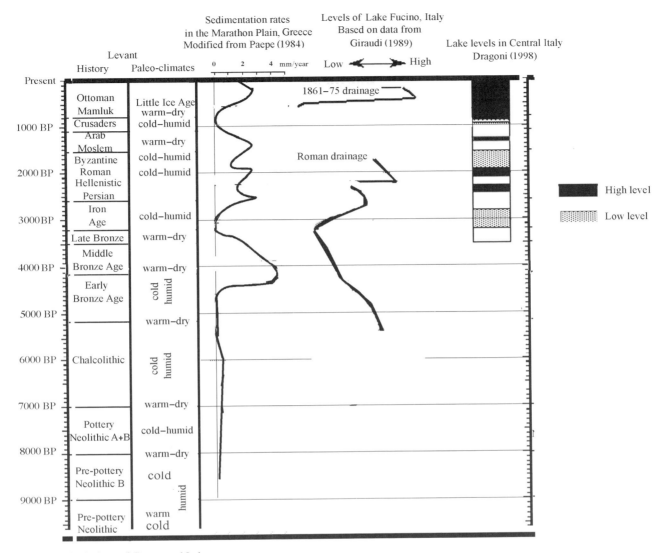

Fig. 2.2. Paleo-hydrology of Greece and Italy.

by low sedimentation, that is, a cold climate. During the Roman period, the rates are high in general, presumably because of deforestation, but one can still see fluctuations that correspond to cooling and warming. During the Moslem–Arab warm spell (*c.* 700 AD), rates of sedimentation became higher, corresponding to the phase of global warming.

The soil grain size median calculated by Harrison *et al.* (1993), who processed published data on lake levels in Greece, indicates an increase in all lake levels throughout the early Holocene, to a maximum between 7.5 ka and 5.5 ka BP with a minimum *c.* 4.5 ka BP. Then there was a return to wetter conditions around 3.0 ka BP, after which there was a general decrease in the levels of the lakes. Comparing these data with the time series of the Levant, the high levels of the lakes in Greece, from 7.5 ka to 5.5 ka BP, may correlate with the cold mid-Neolithic period around 7.5 ka BP and

the cold humid climate of the Chalcolithic (*c.* 5.3 ka to 6.5 ka BP). The warm dry period, with low lake levels in Greece, *c.* 4.5 ka BP may be correlated with the dry warm period of the MB in the Levant, which started *c.* 4.3 ka BP. Yet, neither the cold wet phase of the EB of the Levant (between *c.* 5.2 ka and *c.* 4.3 ka BP) nor the changes observed in the Levant after 2.5 ka BP is observed in the Greek lake levels.

A pronounced stage of destruction and desertion all over Greece, for which archaeologists have found evidence and which caused the collapse of the Mycenaean civilization, occurred *c.* 3.2 ka BP. Yet archaeologists are divided with regard to the reasons for this collapse. Finley (1981, p. 58) tends to connect it with the invasion of Indo-Arian tribes from the north, while Weiss (1982) claims that there is no evidence of newcomers at that time. Also, Carpenter (1966, p. 37) claimed that "These sites were deserted without any trace of destruction or conflagration" and suggested that the destruction of the settlements was a result of severe droughts, which were followed by the migration of the Aegean inhabitants of mainland Greece to the west, north, and east to seek refuge: the Sea People to the Levantine coast and Egypt, the Semitic tribes (the Apiru) to Canaan, and the Hittites to northern Syria below the Taurus mountain rampart. The correlation with the Levant time series supports the climate change explanation for this disaster.

Carpenter (1966) offered a similar argument concerning archaeological evidence of the desertion of many Greek sites in the Peloponnese during the seventh and eighth century AD. He argued against the explanation proposed by some archaeologists that these sites were vacated because of the invasion of Greece by Slavic tribes. According to Carpenter, the Slavs did not reach Greece until two or three centuries later and the exodus of people from their homes was the result of the severely destructive drought, which ended in the ninth century AD, after which "southern Greece was once more fit to support an increasing population" (Carpenter, 1966, p. 79) The time series from the Levant supports this explanation.

2.2.2 The Italian peninsula and the French Mediterranean coast

In general, the climatic regime of the Italian peninsula and French Mediterranean coast is similarly influenced by the westerlies cyclonic system to the north, which dominates the region during winter, and the Sahara anticyclonic zone, which dominates the region during summer months. Thus, this region is characterized by mild, relatively humid winters and hot and dry summers. Sicily and the southern Italian peninsula are areas typical of such a climate. Because of the Apennine mountains, the western half of the peninsula receives more precipitation than does the eastern half. The western coastal area is less cold in the winter ($8\,^{\circ}$C in January), wetter (2000–3000 mm annually) and has fewer instances of freezing weather (in Florence and Rome, the average January temperature is 5–6$\,^{\circ}$C, with 1000 mm annual precipitation) than other areas with this type of climate. To the east, precipitation is much lower (500–600 mm annually) with hot summers (average July temperature is 24–25$\,^{\circ}$C in Calabria and 28$\,^{\circ}$C in Sicily). The dominant summer wind pattern is the Sirocco, originating from North Africa. This wind is searing, dry and violent.

A study of lake levels and climate for the last 30,000 years in the Fucino area central Italy was carried out by Giraudi (1989). This lake is situated in the Apennines, east of Rome. Before being drained, Lake Fucino measured 150 km^2 with a drainage area of 710 km^2. Roughly oval in shape and 19 km long from northwest to southeast, it had a maximum width of 10 km and a maximum depth of roughly 22 m. The surrounding area, essentially of limestone rock, ranged from the 900 m of Monte Salviano

up to the 2349 m of Monte Sirente. Data on variations of lake levels are available on different time scales. For the last period of existence of Lake Fucino, from 1750 to 1861 AD, there are measurements of levels first on a decade scale, then intermittently, then annually and finally in the years before the drainage, monthly. For the period between the Iron Age and the Roman period, there are data for variations over centuries, while for the remaining part of the Holocene, data are known for intervals of a thousand years. There is geological and geomorphologic mapping that has identified delta and lacustrine sediments, as well as shoreline terraces, wave cut terraces and soft-rock pediments. Each provides evidence of ancient levels of this lake during the late Pleistocene and Holocene. The following changes can be recognized (Fig. 2.2):

a slight rise in lake level dating back to an indefinite period before 33 ka BP;

a strong rise in level during the period between c. 30 ka and 20/18 ka BP;

a fall between 20/18 ka and 7.5 ka to 6.5 ka BP;

a rise between 7 ka and 5 ka BP;

a fall between 5 ka and 2.8 ka BP;

a rise between 2.8 ka and 2.3 ka BP, which led the Romans to build a magnificent drainage project;

a fall between 2.3 ka and 1.8 ka BP;

a low level during the sixteenth century AD;

an increase in levels after 1750 AD.

From this, it can be concluded that during the Late Pleistocene, a high level of the lake can be correlated with the advance of the glaciers, most probably because the lake received a higher rate of inflow. This can be explained by the higher ratio of precipitation on its drainage basin. During the Holocene, the picture seems to be more or less the same, namely, that the level of the lake rose during colder periods, while during warm periods it retreated.

Four tree ring series, derived from trees that grew at higher altitude (1750–2300 m elevation) in southern France and southern Italy (Calabria), were used to study summer temperature changes during the period from 1150 AD until the present (Serre-Bachet and Guiot, 1987). A sharp warming occurred around 1705 AD, lasting until about 1810 AD, although with fluctuations and a cold period from 1740 to 1760 AD.

Dragoni (1998) surveyed the historical, archaeological and climatic data for the last 3000 years for the Italian peninsula south of 43° N. He found that there were oscillations from warm climate, equated with dry, to cool, equated with wet, climate and that each successive period lasted a few hundred years. During the warm–dry periods, the level of the lakes was low, whereas it was high during the wet–humid periods. His synthesis of the paleo-lake levels of central Italy showed low levels from c. 1200 BC to c. 750 BC, from 100 AD to 500 AD and from *c.* 1100 AD to 1200 AD. High levels were observed from *c.* 400 BC to 300 BC, from 100 BC to 100 AD, from 750 AD to 850 AD and from 1200 AD to the present. Towards the end of the nineteenth century, a warming of the climate can be observed, together with a reduction in precipitation. This trend continue to the present time. Extrapolating this trend to the next century shows a decrease in the water yield of up to 10–15% compared with that at present (Fig. 2.2).

An interdisciplinary recent investigation of sediments of Lago di Pergusa in central Sicily, the catchment area of which is about 7.5 km², enabled a glimpse at the evolution of the environment of this area during the Holocene (Sadori and Narcisi, 2001). The sequence starts at *c.* 11 ka BP, with indications of the end of the steppe conditions that prevailed during the last glacial period. At *c.* 10.7 ka BP, a process of reforestation started. The climate conditions until 7.5 ka BP were humid, reaching a peak at *c.* 9 ka BP. Sedimentation rates during this period were rather low because of the forest cover. At *c.* 7.4 ka BP, a decrease in the moisture started, which continued to *c.* 4.4 ka BP. Marked dry conditions occurred also between 3.3 ka and 2.8 ka BP, after which more moderate dry conditions occurred. During this period, there was an increase of olives and other agricultural vegetation. From *c.* 3 ka to 2 ka BP, the rate of sedimentation was high because of erosion and the reduced lake level. After that, the sediments do not show any change in climate conditions.

Harrison *et al.* (1993), comparing the differences between levels of lakes in southern and northern Europe during the Holocene, found that the conditions during the early Holocene were as today in southern France but very different to the present in southern Sweden. By comparison, during the middle Holocene, the climatic conditions of these two regions were similar, as the levels of the lakes in both were at their lowest between 4.5 ka and 3.5 ka BP.

Digerfeldt *et al.* (1997) reconstructed the paleo-climates of the Holocene at Haute-Provence, southeast France, according to the paleo-levels of Lac de Saint-Léger. The levels of this lake were high in the beginning of the Holocene. At *c.* 5 ka BP, lake levels began to fall and this continued until 4.5 ka BP. From then until 4 ka BP, there was a moderate rise but soon afterwards the decline continued and reached its maximum between *c.* 3.3 ka and 2.5 ka BP. Later on there was a progressive rise in the level of the lake. The changes of levels were correlated with paleo-climatic data derived from pollen diagrams. The correlation showed that the main changes of lake levels and vegetation were caused by changes of precipitation. From 1.5 ka BP onwards, the changes in vegetation were influenced by human intervention. Another conclusion derived from the climatic interpretation is that high temperatures during the Holocene equated with high precipitation rates in this region (Digerfeldt *et al.*, 1997).

2.2.2.a THE ISLAND OF MALTA

Malta can be regarded as a stepping-stone between Africa and Europe from all aspects, geographical, climatological and anthropological. The climate is typically Mediterranean, namely mild rainy winters and hot-to-warm dry summers. The average annual precipitation is about 600 mm.

Although no paleo-climatic proxy-data are available, the archaeological evidence from Malta may be regarded as proxy-proxy-data. Especially significant is the abrupt and total desertion of the island at *c.* 2500 BC by the people of the Temple period (Trump, 1990). As no evidence was found for destruction by war, it seems reasonable to suggest the warm climate change after the EB as the most reasonable cause for this desertion. Although the impact of this change in the Levant is felt two centuries later, it may be argued that Malta, situated in close proximity to the African continent, felt the impact of this change earlier, as did the city of Arad in the Negev part of Israel, which was deserted at *c.* 2600 BC (Issar and Zohar, 2003).

2.2.3 The Iberian peninsula

Figure 2.3 shows the Iberian peninsula.

2.2.3.a CONTEMPORARY CLIMATE

The climate of the Iberian peninsula is decided by various factors. In the first place, its position on the southwestern flank of Europe, between 36° N and 44° N, causes it to be influenced periodically by the high sub-tropical and the high polar pressure zones. It lies between the Atlantic Ocean and the Mediterranean Sea and is an extension of the Euro-Asian continent; consequently, it forms a passageway for air masses, either of sub-tropical marine and continental (Sahara) or of marine polar and Mediterranean origin (seldom of polar continental origin).

The special mountainous morphology of the peninsula (average altitude of the Spanish Meseta 660 m above mean sea level) also influences the climate. The high Meseta obstructs the free passage of the Atlantic air masses of the westerlies system. Though the valleys of the large transversal rivers enable air masses to penetrate inland, they do not reach the Mediterranean coastal areas.

In general, it can be said that the mountainous character and the size of the peninsula minimize the regulating influence of the seas surrounding the peninsula and thus contribute to its continental type climate, which is distinguished by high and low temperatures during the summer and winter, respectively.

The climate in the southern Iberian peninsula is similar to that found in Sicily, especially in the Andalusia plain. The average temperature is 12–13 °C in January and 27–28 °C in July; annual precipitation is 500 to 700 mm. Along the length of the Mediterranean coast (the "Spanish Levant"), precipitation is less

Fig. 2.3. Map of the Iberian peninsula.

(300–350 mm per year). These characteristics show that this region is a transitional area towards the hot and dry regions of northern Africa (Capel Molina, 1981).

Rodó *et al.* (1997), analyzing data from the Iberian peninsula, Balearic Islands and northern Africa, have found a relationship between the variations in seasonal rainfall during the present century and the North Atlantic oscillation (NAO) and the ENSO. While the NAO influences most of the peninsula except its eastern part, the opposite was found for the ENSO.

2.2.3.b PALEO-GEOMORPHOLOGY

A study of the terraces of the Jarama river, a tributary of the Tajo river, east of Madrid (Alonso and Garzon, 1994) found evidence for a process of river incision in the transitional period between the Late Pleistocene and the Holocene. After the incision stage, a period of stability occurred, long enough to allow the development of a soil of green clay rich in organic matter and the establishment of a forest of phreato-phitic species (such as *Alnus glutinosa* and *Ulmus* spp.). In one sector, the wood was dated and was found to be from 6 ka BP. (This piece of wood may be reworked.) In another sector, the date was 3 ka BP. A conglomerate showing

an erosive base and bearing evidence of multi-episodic infilling, which apparently indicates high and low regimes of flow, overlaid this layer. The conglomerate itself was also overlaid by silts or clays, as well as by lenticular layers of conglomerates and sands. This sequence represents a meandering river of medium-to-low sinuosity, with a bed load composed mainly of gravels. The date of the middle part of the gravel sequence was found to be 2.4 ka BP, while the age of a clay layer from a secondary channel on top of the sequence dates from 0.4 ka BP.

In conclusion, the history of the Jarama river system during the Holocene can be divided into three stages.

1. An incision stage during the transitional Pleistocene–Holocene period;
2. A stable soil-forming phase prior to 6 ka and lasting to 3 ka BP;
3. An alluvial aggradation phase from *c.* 2.5 ka to 0.4 ka BP.

On the basis of similar incision stages in Italy and Greece, Alonso and Garzon (1994) suggest that the incision is a result of human activity, namely rapid forest felling and resultant alluvation. I am more inclined to suggest a climatic reason, as will be discussed below.

Goytre and Garzon (1996) analyzed historical data of floods of the Jucar river, flowing to the Mediterranean south of Valencia, from the fourteenth century to the present. The frequency curve shows a period of high frequency starting at 1720 AD and reaching a maximum at 1800 AD coming to a low c. 1820 AD and then again reaching a peak in 1870 AD and 1890 AD. They correlated the flood frequency and size curve with precipitation data available since 1860 AD. The increase in flood frequency and size from the second half of the eighteenth century till the beginning of the twentieth century also coincided with an increase in winter flooding.

Zazo et al. (1996) analyzed data from studies of sedimentological processes in karstic systems, pollen time series and morphosedimentological studies from the Iberian littoral zone. They concluded that the Mediterranean region had a cold dry climate during the last glacial period, while the Cantabrian region was cold to cool and humid. During the Younger Dryas cold period, the climate was dry but not cold in the Mediterranean region and cool and humid, but less than today, in Cantabria. To the west of the Gibraltar Strait, the climatic conditions were humid and temperate, both during the last glacial maximum and during the Younger Dryas.

The sea level changes along the southern coasts of the Iberian peninsula during the Holocene have been explored rather intensively (Goy et al., 1996; Lario et al., 1995). The main finding is that most of the coasts of southern Spain were influenced by a process of uplift (or sea retreat) during the upper part of the Holocene, which caused spit-bar formation. During the lower part, from 10 ka BP, the sea was rising, reaching its maximum level at c. 6.4 ka BP. This was followed by the retreat of the sea and a prograding spit-bar system developed in the areas from which the sea moved back. The first phase of progradation, starting after the c. 6.4 ka BP maximum, lasted until c. 4.5 ka BP, reaching its maximum at c. 4. ka BP, when a gap in sedimentation occurred. This was followed by a period of retreat of the sea and the formation of spit-bars. A pronounced gap of progradation, accompanied by a pronounced phase of modification of littoral dynamics and evidence for changes of wind direction, from mainly easterlies and north easterlies to westerlies, occurred between 3 ka and 2.75 ka BP. Sea-level retreat and progradation continued until 1.2 ka BP, when there was a gap of sedimentation and a period of sea-level rise, which extended to c. 0.7 ka BP. Estuaries had a greater fluvial than marine influence at c. 1.0 ka BP. From then, a period of retreat continued, with an extraordinary increase in coastal progradation in the littoral zones, reaching its maximum c. 0.5 ka BP.

Two Holocene transgressive phases have been detected along the Cantabrian coast of northern Spain (Altuna et al., 1993). The lower transgression occurred before 5.8 ka BP and the upper one after 4.9 ka BP. During the transgressive phases, coarse material was deposited on the sandy beach material, while during the regression of the sea, coastal dunes and freshwater environments developed. On these flourished a deciduous forest suited to a temperate and humid climate.

A geomorphological study was carried out in the central Ebro valley by Soriano and Calvo (1987). It revealed two periods of intensive deposition, one post-Roman and pre-Visigoth, and the other post-Medieval. The authors suggest that processes of accumulation of gravel were connected with colder climates, while periods of incision associated with warmer climates.

Another investigation of the middle Ebro river system was carried out by Stevenson et al. (1991). This investigation also extended to the saline marshes (saladas) of the region. It was based on pollen and geochemical analyses of cored material from the saladas, as well as geomorphological investigations. Unfortunately, dated samples are scarce. The general conclusions drawn by Stevenson et al. (1991) are that erosional phases conformed with aridization phases and vice versa. They observed three main periods of "cut and fill": a major erosion phase of the Lower and Middle Holocene, followed by an aggradation phase that started with alluvial material and continued with fine-grained and clay material (Cerezuela unit). Unfortunately, Stevenson et al. (1991) give only one ^{14}C date in their section of a hearth found in the mid part of the exposed section, which is 3815 ± 80 BP. On the basis of the similarity in pollen assemblage found at the depth of 150 to 210 cm in Salada Pequenia, they try to make a correlation to the top layers of the exposed section of Cerzuela unit and claim that it is of the Iberian period and is evidence of a humid climate. This correlation, not based on ^{14}C dating, is rather questionable. However, in addition to this interpretation by Stevenson et al., quite a lot of information can be gained from the pollen and chemical data from the outcrop of the Cerzuela unit at Alcaniz. The layers above the dated one, namely after 3.8 ka BP, show an abrupt increase in the pollen of Artemisia spp., Chaenopodiaceae and Gramineae, and a slight increase in Pinus spp., while at the same time, there is an increase in sodium (unfortunately only cations and no anions were analyzed). This may point to a warmer period in which evaporation exceeded precipitation and inflow to the marsh. This agrees with the conclusion from many other investigations, as will be discussed below, that a period of warming up started c. 4 ka BP. The increase in Pinus pollen may suggest that the precipitation in this region did not fall below that which is crucial for the existence of a Pinus habitat, while the warmer climate promoted the expansion of this species.

Evidence for the Holocene glacial episode, attributed to the Little Ice Age, was found in the central and Aragonean Pyrenees (Serrano Candas and Agudo Garrido, 1988).

2.2.3.c PALYNOLOGICAL TIME SERIES

Carrión and Dupré (1996) investigated the Late Quaternary vegetational history at Navarrés, eastern Spain (Fig. 2.4a). The area lies

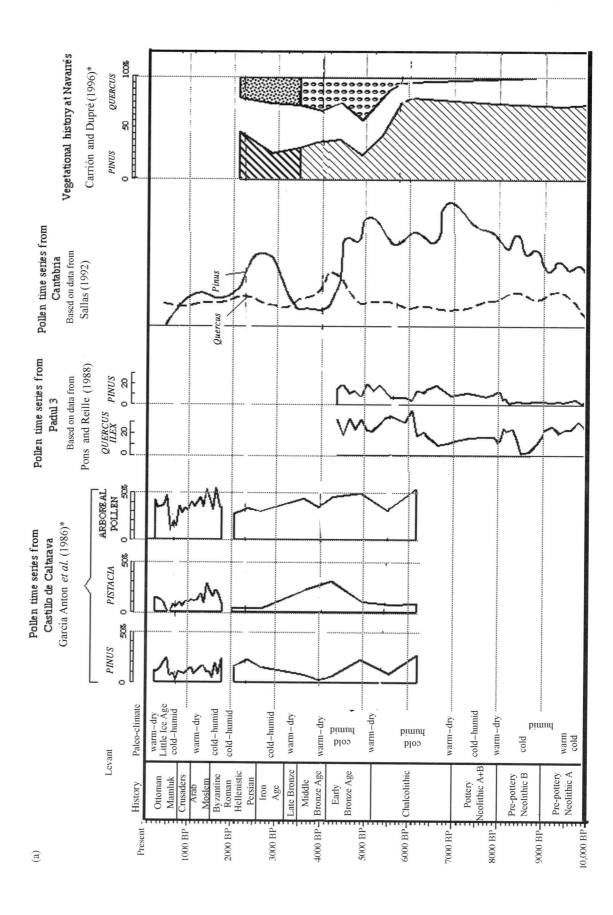

(a)

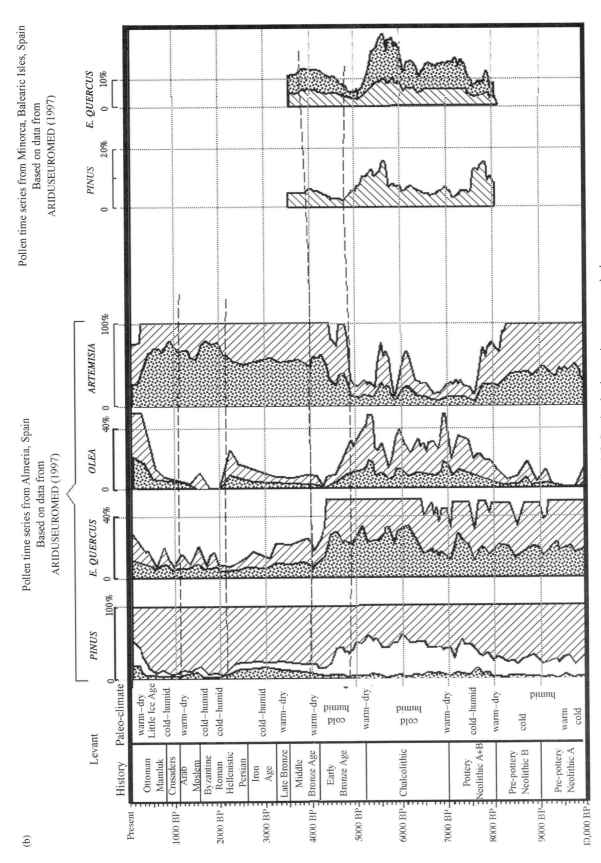

Fig. 2.4. Pollen time series for the Iberian peninsula. * Adjusted to scale and streamlined (3–5) points by the running average method.

(b)

Pollen time series from Minorca, Balearic Isles, Spain
Based on data from
ARIDUSEUROMED (1997)

Pollen time series from Almeria, Spain
Based on data from
ARIDUSEUROMED (1997)

in the lower meso-Mediterranean belt and has a dry to sub-humid climate. The mean annual temperatures are 15–16 °C and mean annual rainfall is 550 mm. In the assemblage dated as Upper Pleistocene, they found the recognized European sequence of upper Pleniglacial, from 16 ka to 15 ka BP, characterized by a high percentage of *Artemisia* spp. (about 19%) while *Pinus* spp. it is only 50%. This means that continental climatic conditions prevailed during the colder periods of the Pleistocene. This period was followed by the Bölling–Allerod warm period, in which the *Artemisia* percentage declined sharply (∼3%) while the *Pinus* percentage increased (to 77–89%). From *c.* 11 ka to 10 ka BP, one can see the influence of the cold Younger Dryas episode, which led to the strengthening of the continental conditions. This is evidenced by an increase in *Artemisia* spp. (up to 14%), Chenopodiaceae and *Ephedra* spp., while *Pinus* spp. decreased (to 67–73%). With the beginning of the Holocene, *c.* 10 ka BP, the climate became warmer. This led to a decline in the percentage of the *Artemisia*, a relative increase in *Pinus* (to 67–84%) and the appearance of evergreen *Quercus* spp. (2–3%). At *c.* 5.5 ka BP, there was a decrease in *Pinus* (to 46–61%) while the *Quercus* spp. increased (reaching up to 14%). At 4 ka BP, *Quercus* reached 32%, while the *Pinus* percentage was 23–53%. This continued until *c.* 3 ka BP.

Carrión and Dupré (1996) discussed the reasons for the expansion of the oaks and their replacement of pines and spruces after *c.* 5 ka BP. They suggest that this change in flora was more a climate change indicator rather than an indication of any impact of human intervention, although, generally speaking, the anthropogenic factor could have had local influence. The data presented by Carrión and Dupré (1996) are shown in Fig. 2.4.

If these authors are correct in believing that climate was the main cause for the changes in the pollen assemblages, one can conclude from the palynological data (even though the dating of the core was sparse) that, *c.* 5.8 ka BP, the climate was cold, getting warmer towards *c.* 5 ka BP and again colder towards 4.5 ka BP. It then warmed up again *c.* 4 ka BP.

Carrión and Dupre (1996) correlated their data with those of Menéndez-Amor and Florschütz (1961), who analyzed a cored section in the same region. Dating was also sparse for this core, but continuing with the same assumption, one can conclude that *c.* 6.2 ka BP there was a rather warm period. Assuming a similar rate of deposition for the lower part of the section as for its upper part, one can conclude that there was a cold period *c.* 7.5 ka BP, while between 8 ka and 9 ka BP it was warm, after a cold period about 10 ka BP.

Another study providing an indicator of climate changes during the lower part of the Holocene in the northeastern part of the Iberian peninsula was by Perez-Obiol and Julia (1994), who studied the pollen record in a core from Lake Banyoles, northeast of Barcelona. The record, which started *c.* 31 ka BP, showed an in-

terstadial event between 30 ka and 27 ka BP, a Pleniglacial Period with minor oscillations that ended abruptly *c.* 14 ka BP. There is also the Younger Dryas event *c.* 12 ka BP. The pollen diagram indicates that the ameliorating climatic conditions led the deciduous *Quercus* spp. to flourish from *c.* 11 ka BP onwards. As *Quercus* flourished so *Pinus* spp. diminished, indicating a climate change. The lack of dating in this study makes detailed conclusions regarding climate changes rather difficult, but by correlating the time scaled section for *Pinus* (calculated by polynomial interpolation on the basis of ^{14}C and U/TH dates) with that of *Quercus*, one can draw very general conclusions with regard to the climate changes in this region during the lower half of the Holocene. This period started with a cold climate stretching from 10 ka to 9 ka BP (with a short warm interval sometime around 9.5 BP), followed by a gradual warming up, which reached its maximum around 8 ka BP. A profound cold period, when *Quercus* falls and *Pinus* surges, reached its maximum around 6 ka BP, with an intermediate warm phase *c.* 5 ka BP.

Pons and Reille (1988) presented a detailed picture of the impact of climate changes on the natural environment in the southern part of Spain for the period from before the last glacial period to the mid Holocene (Fig. 2.4a). The site of the borings is a peat bog at Padul, situated in a long and vast tectonic valley at the eastern foot of the Sierra Nevada. The bog was most probably fed rather constantly by groundwater. The Oldest Dryas, *c.* 15 ka BP, was marked by an increase in steppe species (*Juniperus*) at the expense of *Pinus*, which indicates dry continental conditions. This was followed by a warmer period when *Quercus* and *Pistachia* spp. became more abundant. The Younger Dryas cold phase (between dated layers of 12 ka and 10 ka BP, with a hiatus in between) was clearly characterized by a reduction in *Quercus* and *Pistachia* and an increase in *Pinus*. *Artemisia* and *Juniperus* also become more abundant: evidence for steppe conditions. At *c.* 10 ka BP, the warmer climate brought a recurrence of *Quercus ilex* and *Pistachia*, while *Pinus* decreased. At *c.* 8.5 ka BP, there was again a decrease in *Pinus*, but at the same time there was also a decrease in *Quercus* spp. and some decrease in *Pistachia*, in the total arboreal flora and in the Poaceae. By comparison, fern spores showed an abrupt increase, while *Juniperus* and *Artemisia* spp. practically disappeared. Although Pons and Reille (1988) considered this to be of local significance, I would suggest the possibility of some kind of a climatic anomaly, for example higher temperatures and the strengthening of summer precipitation. Between *c.* 8.2 ka and 8 ka BP, there is an abundance of *Quercus* with the almost total absence of steppe species, which suggested to Pons and Reille "... that optimal postglacial thermic and humid conditions were then prevailing". The reduction in *Q. ilex*, *Q. suber* and *Pistachia*, together with the increase in *Pinus*, was seen by Pons and Reille as reflecting conditions of more open regional forest formation, while I consider it as a sign of a colder phase. Just after 8 ka BP, there was another increase in *Quercus* spp., *Pistachia* and *Olea*, while *Pinus* slightly decreased, which implies a Mediterranean climate. During the period immediately following, 7.5 ka to 6.5 ka BP, *Olea*, *Pistachia* and evergreen *Quercus* spp. decreased, while *Pinus* was more abundant, which may betoken a colder climate (although deciduous oak increased). At 6.5 ka BP, *Olea* returned, *Quercus* and *Pistachia* increased, while *Pinus* decreased. This trend reversed *c.* 6 ka BP.

Changes in the ratio of *Quercus* and *Pistachia* to *Pinus* can also be found in the palynological time series from cores at Castillo de Caltarava, in the sediments in the marshes close to the Guadiana river (Garcia Anton *et al.*, 1986; Fig. 2.4a). In the general section, one can see that there are changes in the ratio of *Pinus* to *Pistachia*. These changes are climatic indicators because, as shown by Pons and Reille (1986), high comparative levels of *Pinus* are characteristic for the glacial periods, and vice versa, for the Granada region. The scarcity of ^{14}C dating does not permit a good correlation with the other curves, but from the dates available, one can see that *c.* 6.2 ka BP, there was a high ratio of *Pinus* to *Pistachia*, which might correspond to a colder climate. This was followed by a short period of decrease in the ratio, which may correspond to a warmer phase. At *c.* 5 ka BP, there was an increase in the ratio of *Pinus* to *Pistachia*, which may represent a cold and humid period, and at *c.* 4 ka BP, there was a strong decrease in *Pinus* and increase in *Pistachia*, which may represent a warm and dry period. Approaching the layer dated *c.* 1.7 ka BP, there was another increase in *Pinus*, which was interrupted about 2 ka BP (not dated) by a layer of about 10 cm devoid of pollen. I would suggest that this is equivalent to the period of high precipitation during the early Roman period in which the marshes turned into a flowing river, which caused the oxidation of the sediments. At *c.* 0.5 ka BP, there was an increase in *Pinus* and reduction in *Pistachia*, which most probably can be correlated with the Little Ice Age.

Dupre *et al.* (1996) reconstructed the paleo-environment of the lake of San Benito in central eastern Spain, near Valencia. They found a clear predominance of pines during the Upper Pleistocene when lagoonal conditions prevailed. Up to the Middle Holocene, fluvio-aluvial conditions prevailed. At the top of the section, *c.* 1.4 ka BP, the pollen profile showed the regeneration of the tree cover, mainly *Quercus*, which the authors believed was the result of the decline of intensive agriculture after the fall of the Roman Empire.

Ruiz Zapata *et al.* (1996) reconstructed the paleo-climates of the central western part of the Iberian peninsula from an east–west cross section based on four cores drilled in intra-mountain depressions. Two of the cores were drilled in glacio-lacustrine deposits, while two comprised periglacial deposits. Pollen assemblages for the central zone showed that, between 8 ka and 7 ka BP, *Betula* showed an increasing trend, while *Pinus* was more or less stable and *Quercus* diminished, disappearing *c.* 7 ka BP. *Quercus*

reappeared *c.* 4.5 ka BP. At this time, *Pinus* reached a minimum, while *Betula* reached a peak and from then on diminished, until it disappeared altogether at 2.3 ka BP. At 4.5 ka BP, *Pinus* was at a minimum; it reached a peak at *c.* 3.2 ka BP, with a certain minimum at 1.7 ka BP and another between 0.8 ka and 0.5 ka BP. *Olea* appeared *c.* 2.5 ka BP, disappeared *c.* 2 ka BP and reappeared *c.* 1.7 ka BP, to disappear again for a short while between 1.5 ka and 1.3 ka BP. From this period on, *Olea* spp. increased. In the eastern part of the cross section, *Pinus* spp. were at a maximum at *c.* 2 ka BP, decreased *c.* 1.3 ka BP, showed a short peak *c.* 1.2 ka BP and slowly decreased to a minimum *c.* 0.5 ka and *c.* 0.25 ka BP. By comparison, *Quercus* spp. were at a minimum in the eastern part of the section from 1.4 ka to *c.* 1 ka BP and reached a maximum *c.* 0.7 ka BP.

A multidiscipline investigation was carried out on the deposits in a cave near the seashore on the central eastern coastline of Spain (Badal *et al.*, 1993). The sediments, about 3.5 m thick, contained artifacts of the Iberian Neolithic and Bronze Ages. The lowermost layer was dated at 7540 ± 140 BP, but this seems to be too early a date as the archaeological remains of Ceramic Neolithic I A, are only 7 ka years old. The other ^{14}C dates more or less corresponded with the archaeological findings. The pollen assemblage was rather poor in arboreal and rich in non-arboreal taxa. The major tree component was *Pinus*, followed by *Quercus* (*Q. ilex* and *Q. faginea*). The interchange between these trees seems to be a climatic indicator, according to Dupre *et al.* (1996), who divided the section into four zones. Zone A (350–260 cm) contained little tree pollen, mainly *Pinus*, and was considered as created during a dry period. Zone B (255–185 cm) was rich in *Quercus* pollen and was considered to be the most humid part of the sequence. Zone C (175–115 cm), in which *Pinus* replaces *Quercus*, was considered to be influenced by human factors. However, the same reason was given for the reciprocal trend in zone D, where *Quercus* replaces *Pinus*. As already discussed, I am more inclined to believe that the interchange between *Pinus* and *Quercus* (and *Pistachia*) is a reaction to climate change, rather than reflecting anthropogenic impact.

As past of a multidiscipline program (ARIDUSEUROMED, 1997), a group of Spanish palynologists have investigated the pollen paleo-records for the coast of Almeria, in southeastern Spain, which is semi-arid, with an annual mean precipitation of *c.* 250–350 mm and mean annual temperature of 18–21 °C.

In a core from the salt marsh of San Rafael in this region, the layers were mostly clay to 10.75 m, peat from 10.75 to 11.75 m, clay at 11.75–14.50 m, argillaceous slime 14.50–18.00 m and gravel at the bottom layer to the depth of 19.00 m. The group studying this core interpreted the pollen assemblages in the following way.

Layers deposited between 18 ka and 15 ka BP (bottom to 18.00 m) contained mainly arboreal pollen, composed of *Olea*, deciduous and non-deciduous *Quercus* and *Pinus*, evidence of a relatively warm and humid environment. From 15 ka to 7 ka BP (18.00–13.50 m), the assemblage showed a decline in arboreal pollen and an increase in steppe type vegetation. There are indications that during this period there was a decrease in temperature as well as a possible decrease or change in the distribution of the precipitation. From 7 ka to 4.5 ka BP (13.50–6.50 m), there was a decline in steppe vegetation together with an increase in arboreal pollen and certain shrub taxa. This shows a climate optimum, most probably warm and humid. From 4.5 ka BP upwards, there were indications of the encroachments of arid conditions. This was evidenced by an increase in steppe vegetation and a decrease in arboreal pollen. Deciduous *Quercus* disappeared and evergreen *Quercus* and *Olea* declined. The uppermost part of the core, presumably from c. 0.1 ka to 0.2 ka BP, showed an increase in arboreal pollen, including *Pinus*, *Quercus* and *Olea*, and a decrease in *Artemisia*, which may be a result of anthropogenic as well as natural processes.

The investigation of the paleo-biotic assemblage and lithology of cored sediments from the bottom of Laguna de Medina, near Cádiz in southwestern Spain, gave a detailed paleo-environmental history of the lower part of the Holocene (Reed *et al.*, 2001). These cores suggest that there was a dry phase from c. 8 ka to 7.2 ka BP and a humid period from c. 6.9 ka to c. 6.7 ka BP (calibrated (cal.): [14]C dates given as "absolute" dates by calibrating either locally, with varved lake deposits or tree rings, or generally, with a general calibration curve).

Based on climate changes observed in other parts of the Mediterranean, and considering that the climate of the Almeria coast is Mediterranean, I would suggest a somewhat different conclusion with regard to the climate changes during the Holocene. It is possible that the peak in steppe flora, especially *Artemisia*, and the strong decrease in arboreal flora, especially *Pinus* and deciduous *Quercus*, between c. 10 ka and 7 ka BP was not caused by a decrease in temperature but rather by a relatively warm and dry period. When there was an increase in arboreal flora and decrease in steppe flora (7 ka to c. 4.5 ka BP), the region enjoyed a mainly cooler and more humid climate. This interpretation is supported by the accumulation of peat layers and higher charcoal percentage in the layers deposited during this period. At c. 4.5 BP, there was a considerable increase in temperature, which caused another increase in the steppe flora and decrease in arboreal flora.

In the same report (ARIDUSEUROMED, 1997), one can also find palynological data from the island of Minorca. The climate of the southern part of the island, where the core was taken, is Mediterranean semi-arid, with an annual precipitation of 450 mm. The four to five summer months are dry. The bottom of the section dated to c. 8 ka BP and the floral assemblage was characterized by the dominance of *Buxus* and *Corylus* (taxa that are absent today in Minorca) and high values of *Juniperus*, *Ephedra* and *Quercus* (deciduous and evergreen). High values of *Typha* indicated the

presence of coastal marshes. At c. 6 ka BP, there was a pronounced change: *Olea* appeared while *Buxus* and *Corylus* declined. Some time before 5 ka BP, *Buxus* and *Juniperus* disappeared, while *Olea* increased to reach a peak c. 4.5 ka BP, after which it decreased a little. In the uppermost zone, starting at 4.5 ka BP, *Pistachia* increased, and there was a rise in Chenopodiaceae, which is evidence for the spread of coastal marshes, most probably because of a rise in sea level.

A diagram summarizing the total arboreal pollen frequency of the Minorcan site shows a rather low frequency in the lower part, from c. 8 ka to c. 7 ka BP, followed by an increase between c. 6 ka and 5 ka BP and a strong decrease c. 4.5 ka BP.

A rather recent palynological study in south-central Spain (Carrión *et al.*, 2001) indicated the ecological changes during the Holocene over a region that spreads along the boundary between semi-arid plateau and mountain environments. *Pinus* dominated from c. 9.7 ka to 7.5 ka BP (cal.), which the authors believed was a consequence of the relatively dry climate and natural fires. From c. 7.5 ka to 5.9 ka BP (cal.), there was a moderate invasion by *Quercus* as a result of increasing moisture and temperature. From c. 5.9 ka to 5.0 ka BP (cal), *Pinus* was replaced by deciduous *Quercus*, as well as by *Corylus*, *Betula*, and *Alnus*, etc. From c. 5.0 ka to 1.9 ka BP (cal), the Mediterranean type of forest including *Artemisia* dominated. From c. 1.9 ka to 1.1 ka BP, *Pinus* became dominant and from 1.6 ka BP onwards, human impact became influential. Although it is difficult to correlate this section with the Mediterranean climate change timetable because each one of the intervals expands over more than one division of the Holocene, the impact of climate and not human activity on the vegetation during most of the Holocene is stressed by the authors.

Yll *et al.* (1995) produced a synthesis of the history of the vegetation landscape of the eastern part of the Iberian peninsula and Balearic Islands during the Holocene. They found two fundamental points of climatic impact on the natural environments that defined periods of accentuation of aridity. These occurred at 6 ka and at 4 ka BP.

Ruiz Bustos (1995) investigated the climatic conditions during the Last Ice Age by comparing the mammal remains from this age with the mammal population of the present. His main conclusion was that the Spanish climate was cold and dry during the Würm. He cites Riquelem Cantal (1994), who investigated eight Bronze Age sites in southern Spain dating from the lower half of the second millennia BC (i.e., between 5 ka and 4.5 ka BP, equivalent to the Early Bronze Age of the Levant). The faunal assemblage indicated a cold and dry climate during this period.

Information with regard to changes in aridity during the Holocene has been obtained through the investigation of ^{13}C/^{12}C ratios in grain cereals collected in archaeological sites in Catalonia (northeastern Spain) and Andalusia (southeastern Spain) (Araus

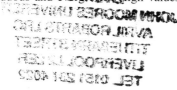

et al., 1997). Lower values of d^{12}C ‰, which show higher water use efficiency, resulting from reduced water supply and/or increased temperature, were consistently found in eastern Andalusia than in Catalonia throughout the period ranging from Neolithic to Iron Age. According to these authors, this shows that Andalusia was drier throughout this period. The data for Catalonia were available from *c.* 6.5 ka BP (Iberian Neolithic) and showed a gradual decrease of values of d^{12}C ‰ and a tendency to more arid conditions extending through the periods for which samples were available: 4.2 ka BP (Iberian Chalcolithic Bronze) to 2.5 ka BP (Iberian Iron Age). There is a small increase at 0.8 ka BP, and a strong decrease towards the present.

Sallas (1992) used a pollen time series in Cantabria to analyse climate changes during the Holocene (Fig. 2.4a). He distinguished three main climatic phases: a cold and dry period from 10.2 ka to 7 ka BP, a warm dry period from 7 ka to 5 ka BP and a colder period from 5 ka to the present. Using the *Pinus* to *Quercus* ratios as a climate change indicator, one can suggest a more detailed division of the Holocene based on his data than that suggested by Sallas (1992; see his Fig. 5). This more detailed division should not be overemphasized as control of dating intervals is lacking. However, it is worthwhile noting that the highest ratios of *Pinus* to *Quercus* at *c.* 7 ka BP and at 5 ka BP may be evidence for a colder climate. Sallas's data appear to indicate a warm phase starting *c.* 4.5 ka BP and continuing until *c.* 3.5 ka BP, followed by a colder phase from *c.* 3 ka BP to 2.3 ka BP.

An evaluation of the changes in precipitation during the last 20 ka years was carried out by Igor Parra (1994), who investigated the pollen assemblages and oxygen isotopes of two marine cores, one in the Mediterranean off the coast of southeast Spain (SU 8103) and the other in the Atlantic (SU8113) off the southern coast of Spain, northeast of the Straits of Gibraltar, and two continental cores, one on the northeastern coast of Catalonia and the other on the western part of the island of Majorca. The palynological analysis of the cores was based mainly on the variations of the four species of *Quercus* and on the comparison of the abundance of each of these types with other pollens, mainly *Pinus*, *Cedrus*, *Artemisia* and *Ephedra*. Data from the data bank of pollen carried by air between Catalonia and Andalusia was used to determine the ratio of pollen as a function of climate and precipitation. The ratios of *Pinus* to *Quercus* were mainly used. These findings, correlated with the data from the cores, gave rise to the following conclusions. According to the Atlantic core, the most humid periods were at 12 ka, 9 ka and 3 ka BP, while according to the Mediterranean core, they occurred *c.* 12 ka, 9 ka and 7 ka BP. The continental cores indicated that the most humid periods were *c.* 5 ka and 3 ka BP, while there was a pronounced period of aridization around 4 ka BP. Parra (1994) recognized an anthropogenic influence from *c.* 6 ka BP but maintained that it did not influence the imprint of climate change on the pollen ratios.

By comparison Van den Brink and Janssen (1985) did interpret the pollen data obtained from a core of a small pond at an altitude of 1600 m in the Serra de Estrela in Portugal in anthropogenic terms. However, they did find that the *Quercus–Betula* forest was destroyed by fire at 4.3 ka BP and replaced by heath, while *c.* 3.2 ka BP, there was a temporary regeneration of *Betula*, corresponding to climate changes suggested by other authors at other sites. It is possible, therefore, that their anthropogenic conclusions are overemphasized. The recent destruction of the forest at 0.85 ka BP may well have been connected with human activities.

A more balanced approach, with regard to the influence of climate change versus human activity for the same region, is demonstrated by van der Knaap and van Leeuwen (1995). They found the following five stages.

1. From *c.* 10.4 ka to 8.7 ka BP, the late glacial steppe changed into a xerothermic forest under warm and rather dry conditions.

2. From *c.* 8.7 ka to 5.7 ka BP, the climate became moister and cooler, and from *c.* 8.2 ka BP, the forest changed from xerothermic to mesothermic. The anthropogenic activity started to show but played a minor role.

3. From *c.* 5.7 to 3.2 BP, the area covered by forests was hardly affected by human activity, yet local overgrazing with soil erosion started *c.* 4.5 ka BP.

4. From *c.* 3.2 ka to 1 ka BP, large-scale deforestation occurred, with regeneration phases in response to human activity.

5. From 1 ka BP to the present, anthropogenic activities have caused the forest to disappear. Human activity is claimed to be the cause of the influx of pine pollens.

Although this analysis is less anthropogenic than that of Van den Brink and Janssen (1985), it is nevertheless likely that climatic, rather than anthropogenic, reasons should be considered for the deforestation *c.* 4.5 ka BP (i.e., a warm period) and for the pine increase *c.* 0.5 BP (Little Ice Age).

Although it is clear that much better chrono-stratigraphy of the Holocene of the Iberian peninsula would have been obtained with more ^{14}C and other types of dating, some conclusions can still be reached by following well-established guide stratigraphic horizons.

One clearly demarkated phase in various pollen time series is the chronozone of *c.* 4 ka BP. The sections in Padul 3 (Pons and Reille, 1988; Fig. 2.4.a) and Cala en Porter in Minorca (ARIDUSEU-ROMED, 1997; Fig. 2.4b) terminate at this period, while the profile in San Rafael in Almeria (ARIDUSEUROMED, 1997; Fig. 2.4b) shows an abrupt reduction in *Quercus* and *Olea*, and an increase in *Artemisia*. In the section at Castillo de Caltrava (Garcia Anton *et al.,* 1986; Fig. 2.4a), one can see a reduction in *Pinus* and total arboreal pollen, with a relative increase of *Quercus*.

Similarly, the profile in Cantabria (Sallas, 1992; Fig. 2.4.a) shows an abrupt decrease of *Pinus* and a certain increase in *Quercus*. Investigation of the middle Ebro river system (Stevenson *et al.*, 1991) showed that, in the layers immediately after 3.8 ka BP, there was an abrupt increase in the pollen of *Artemisia*, Chaenopodiaceae and Gramineae, and a slight increase in *Pinus*. At the same time, there was an increase in the sodium content of the salina, all of which most probably indicate a warmer period.

Information on special Christian ceremonies shows that during the years 1675 to 1715 AD, namely Late Maunder Minimum, which belongs to the Little Ice Age, there were prayers regarding too much precipitation; that is, there were too frequent passages of low-pressure systems over the peninsula (Barriendos, 1997).

2.2.3.d REGIONAL CORRELATION

Geomorphological data show sea-level changes along the southern coasts of the Iberian peninsula. The first phase of progradation, starting after the *c.* 6.5 ka BP pollen maximum, lasted until *c.* 4.5 ka BP and reached its maximum at *c.* 4 ka BP, when a gap in sedimentation occurred (Goy *et al.*, 1996).

One may conclude that there was indeed a severe change to a warmer climate, spelling dryness, at 4 ka BP, at least in the southeastern part of the Iberian Peninsula. This change was global, as it coincided with a high sea level.

This is also the opinion of Jose S. Carrión Garcia (personal communication, 1997), who thinks that increased dryness could have provoked many of the peat sites to stop peat formation. This possibility was also noticed in the thesis by Parra (1994). Perhaps the Azores high pressures had moved latitudinal and, while the southwest was still affected by the southwesterly storms, their influence in the southeast became negligible.

If 4 ka BP is taken as a correlation chronozone, then, despite the scarcity of dating, the re-examination of the Iberian time series suggests that the three millennia prior to this chronozone contained longer colder and less-arid periods. Such a climate was also seen *c.* 3 ka BP, that is, after the warm period that started *c.* 4 ka BP came to an end. One can also claim that the climate was cooler *c.* 2 ka BP, as the section at Castillo de Caltrava (Garcia Anton *et al.*, 1986; Fig. 2.4a) showed an increase in *Pinus* and decrease in *Pistachia*. As mentioned earlier, the influence of the Little Ice Age can also be discerned. A more detailed stratigraphy of the Holocene of the Iberian peninsula cannot be attempted until more dated profiles are available. In general, it can be said that the changing ratios of pollen, as can be seen in the rather detailed pollen time series, indicate many more changes in climate than those envisaged by the Blytt–Sernander division of the Holocene. However, much more detailed dating is required before a new division of the Holocene in the Iberian Peninsula could be considered.

Correlating the impact of the climate changes during the Holocene along the western shores of the Mediterranean (Goy *et al.*, 1996) with those to the east, one can see that the influence of the most pronounced changes was similar. Thus, during the part of the Neolithic when the climate was warm, the sea level came up, while during the Chalcolithic and EB, it mainly receded. The MB warm period was marked by an ingression of the sea, which reached a climax at 4 ka BP. During the Lower Iron Age, which was cold, there was again a regression, the same being true for the Roman period and the Little Ice Age. However, during medieval times (the Arab period, which was warm), the sea advanced inland.

The question, which has to be answered now is, what was the impact of these changes on the hydrological cycle? To answer this question, it is first important to confirm that the changes observed in the time series are indeed a result of climate changes, rather than anthropogenic. Second, it is important to confirm the climatic interpretation of the palynological time series with regard to changes in temperatures and humidity, for example the pine/oak ratio, and then its chronological correlation with the geomorphological observations. Needless to say, the scarcity of dates in the profiles in the Iberian peninsula makes this rather difficult and also makes it hard to draw parallels with the continental section of the Levant for verification. Taking these problems as constraints rather than obstacles, it is suggested that the most obvious key horizon be chosen, in this case the 4 ka BP chronozone, which was prominent in both the Iberian peninsula and the Levant. As this was a conspicuously warm period, one can draw conclusions about its impact on the hydrological cycle and, in general, one can conclude that opposite phenomena occurred during a cold period. As most of the geomorphological data discussed in this chapter lead to the conclusion that there is evidence of a drier climate *c.* 4 ka BP, one can reach the general conclusion that warm periods during the Holocene had a negative effect on the hydrological cycle, while a colder climate brought more rains and humidity. It should be noted that there was a difference between the climate of the Pleistocene and that of the Holocene. In the Pleistocene, most of the Iberian peninsula was cold and dry during glacial periods, while interglacial warm periods were more humid.

Ayala-Carcedo and Iglesiaz López (1997) used a general circulation model to investigate the possible impact of climate change on the hydrological resources of Spain for the year 2060, taking into account a predicted rise of 2.5 °C in the average annual temperature. They estimated that precipitation will be reduced by 6–8% in the northern part of Spain (except the most northwestern provinces of Calicia and the Basque country, where it will be reduced by 2%) and 9–17% in the southern part. In general, this will reduce the surface and subsurface hydrological resources. The change will be affected by a high degree of annual variability. The major impact will be in the southern part of Spain.

Consequently, in spite of the scarcity of precise dates for past samples, it can be concluded that during the past, and, therefore, future warm periods, the south and especially the southwestern part of the Iberian peninsula would become much drier, while other areas might become somewhat drier.

2.3 THE ALPS

From the climatic point of view, the Alps form the belt that divides the Mediterranean regime from that of central Europe. In addition to this general attribute, special local features are governed by the altitudes of the various chains and by their bearings. Records of temperature changes in the Alps, as derived from dendrochronological studies, are available from 8 ka BP to the present (Bircher, 1986). Although the number of oscillations and their range were quite evenly distributed during the Holocene, an extremely cold event can be discerned c. 3.3 ka BP. The last millennium, except its final two centuries, was, on average, colder: the Little Ice Age. Yet one should take into consideration the fact that the measurement device, which in the case of dendrochronological studies is the trees, may not detect or may obscure extremities of climatic conditions.

Jus (1982) investigated the levels of Swiss midland lakes and found a correlation between high levels and wet and/or cold periods, and vice versa. Thus, the period between c. 7.5 ka and c. 6.4 ka BP was one of high levels and it was followed by a period of low levels that lasted to 5.5 ka BP. A period of intermediate levels occurred between c. 5.5 ka and 5.2 ka before lake levels retreated back to a low level until 4.4 ka BP. Then came a short period of high levels until c. 4.3 ka BP, which was followed by another period of low levels until 3.7 ka BP. From 3.7 ka to 3.1 ka BP, levels were high, and this was followed by a period of low levels (3.1 ka to c. 2.8 ka BP) and then one of high levels (c. 2.8 ka to c. 2.5 ka BP, the end of the surveyed period).

The history of glaciation in the Swiss Alps was investigated by a study of the sediment column cored into the bottom of Lake Silvaplane, a proglacial lake (Leeman and Niessen, 1994). Glacial varves (indicative of glacial conditions) were deposited in the lake until 9.4 ka BP and deposition then stopped until 3.3 ka BP, when it restarted. The maximum thickness of varves, denoting maximal glaciation in the catchment area, occurred from 1790 to 1870 AD. According to Leeman and Niessen (1994) the changes in the glaciers' thickness reflected relative changes of summer temperatures rather than changes in the rates of precipitation.

Haas et al. (1997), investigating lake sediments on the Swiss plateau and timberline fluctuations in the Alps found eight synchronous cold phases in pre-Roman times: 9.6 ka to 9.2 ka, 8.6 ka to 8.15 ka, 7.55 ka to 6.9 ka, 6.6 ka to 6.2 ka, 5.3 ka to 4.9 ka, 4.6 ka to 4.4 ka, 3.5 ka to 3.2 ka, and 2.6 to 2.35 ka BP.

The fluctuations in lake levels in the Jura and the sub-Alpine mountain ranges of France during the Holocene were investigated by Magny (1992) and others. The investigations included lithological analysis of the sediments of the lakes. These were correlated with palynological, dendrochronological and archaeological investigations, as well as with radiocarbon dating carried out in this region. Magny also tried to correlate his results with the stratigraphical divisions of Blytt–Sernander but, as could have been predicted, the number of fluctuations he observed was greater than the number of divisions suggested by Blytt–Sernander, especially in the upper part of the section. Magny found similar fluctuations in the Holocene lake levels for a large number of the lakes in this region and concluded that these fluctuations were caused by climatic changes during the Holocene. Correlating his diagrams with the paleo-climates of the Levant (Fig. 2.5), it can be seen that after the Neolithic period, during most of the cold and humid periods in the Levant, the lake levels in the Alps were high. This is especially apparent during the dry chronozone of 4 ka BP, which was discussed in Section 2.1.4.d.

The position of morainic deposits suggests that glaciers advanced during the Little Ice Age, from c. 1590 to 1850 AD (Le Roy Ladurie, 1971). A correlation between the extension of the glaciers in the Alps since 1818 AD and climatic records exists for the Great St. Bernard Pass on the Swiss–Italian border. It was found that glacial advances occurred after a few decades of high precipitation, during the period of accumulation of ice, when there are low temperatures during the seasons of ice ablation (Porter, 1981).

2.4 WESTERN AND NORTHWESTERN EUROPE

The impact of climate changes on lake levels in Europe during the Holocene was investigated using a water balance model (Harrison et al., 1993) in which the relation between the changes in runoff and evaporation could be simulated. Into this model, various constraints were introduced, including insolation anomalies deduced from orbital variations, temperature anomalies inferred from pollen analysis, and cloudiness anomalies deduced from changes in the position of the sub-tropical anticyclone. The simulations of the model show that precipitation was the main factor responsible for changing the water balance, thus affecting lake levels, but that evapo-transpiration did not play an important role in runoff. The data show a difference between northern and southern Europe with regard to the behavior of the lake levels at 9 ka BP. In northern Europe, the levels are low (southern Sweden and Estonia) while they are high in southern Europe. In order to explain this difference, Harrison et al. (1993) suggest a northward shift of the southern tropical anticyclone. This is similar to the

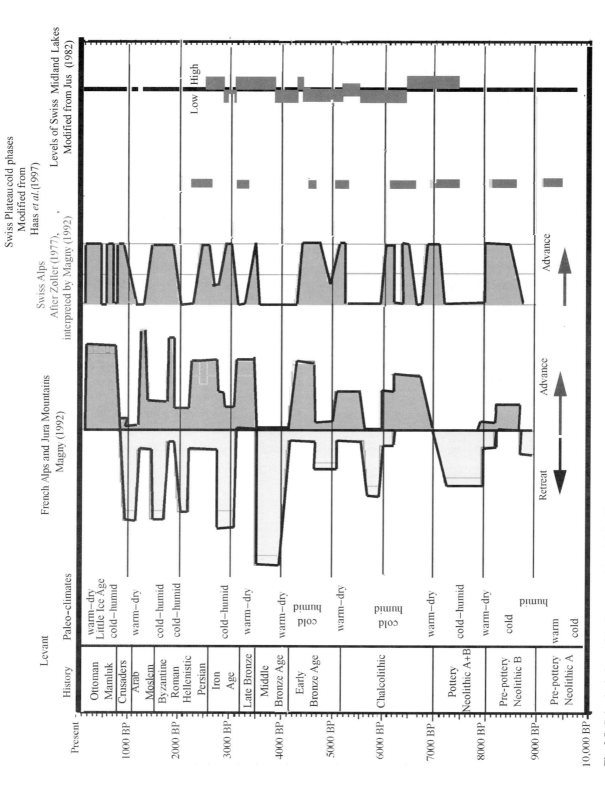

Fig. 2.5. Paleo-hydrology of the Jura mountains and the Alps.

explanations of Kutzbach and Guetter (1986) and Hastenrath and Kutzbach (1983) with regard to variations in lake levels in east Africa during the Holocene. It is also in agreement with Issar's conclusion (1990) that the climate in the Levant during the PPN B was warm and humid.

Warm periods during the Lower Holocene coincided more or less with the Boreal and Atlantic stages of the classification by Blytt–Sernander (Blytt, 1876, Sernander, 1894), for the western European Holocene, which are:

Boreal period: warm and dry, 9.5 ka to 7 ka BP;
Atlantic period: warm and wet, 7 ka to 5 ka BP.

A thermal optimum in Europe, from 8 ka to 5 ka BP, was also identified from the extension of plants and animals north of their present climatic limits, for example the water chestnut and the pond tortoise (Iversen, 1973; Roberts, 1989). On the basis of evidence from fossil beetles (Coleoptera), studied by Coope (1975), temperatures in Britain rose sharply after the cold Younger Dryas stadial at c. 10 ka BP.

The interconnection between climate changes during the Late Quaternary and biotic changes in the terrestrial and lacustrine environments of northwestern Europe was examined by Birks (1986). After the ice glacial conditions reached a peak at c. 18 ka BP, warmer and more humid conditions followed. At c. 14 ka BP, an abrupt amelioration took place. Lakes became abundant in the deglaciated regions from the melting of ice blocks. By c. 1.5 ka BP, the climate deteriorated, as summer and/or winter temperatures decreased; this was accompanied by decreased snow amounts and by droughts. From c. 13 ka to 11 ka BP, a period of strong climate fluctuations occurred. From 11 ka to 10 ka BP (Younger Dryas), temperatures decreased, the retreat of Scandinavian and British Isles ice was halted and the ice cover even regained ground. Woodlands became sparse or even disappeared and were replaced by heaths. The landscape became unstable with eolian activity, solifluction and redevelopment of ground-ice formation and extensive flow of minerals into the lakes. The Younger Dryas cold period ended abruptly, and from 10.3 ka to 10 ka BP warm conditions prevailed, bringing renovation of soil-producing processes. Temperate conditions allowed deciduous forest to prevail from 9 ka to 4 ka BP. From 5 ka BP onwards, the biotic environment, according to Birk's (1986) synthesis, was mainly a function of human interference.

2.4.1 The Netherlands

2.4.1.a CONTEMPORARY CLIMATE
The contemporary climate of the Netherlands is characterized by its latitudinal location within the zone of the westerlies cyclonic system, and by its close proximity to the Atlantic Ocean, which causes its climate to be rather temperate especially in its western

part, where frost is rare. The average annual temperature is 8.5 °C in the north, rising to 10 °C in the south. The average annual precipitation is 720 mm and spring is the driest season.

2.4.1.b CLIMATE CHANGES DURING THE UPPER PLEISTOCENE
Zagwijn (1974) in his summary of the paleo-geographical evolution of the Netherlands during the Quaternary describes the conditions during the last glacial period as being equivalent to those of the present permafrost zone. The regular occurrence of frost wedges in the sediment from this period provides evidence of such permafrost soil conditions. During the Quaternary Period, huge quantities of sediments were deposited. The forest vegetation of the Eemian interglacial gave way to open tundra and a polar desert environment. The majority of sediments deposited during permafrost conditions were eolian sands. These sands are called cover sands, since they cover all the underlying morphology. Other permafrost deposits were sediments carried by melted water from slope solifluction, as well as lacustrine sediments in lakes formed by the impermeable permafrost conditions. The elevated areas of the region were severely eroded. At the end of the Weichselian glacial period, about 10.5 ka BP, the sea level was still low and the greater part of the North Sea was land. As the climate ameliorated, the sea level rose and invaded the North Sea area. It eventually reached the coastal area of the Netherlands. Three zones of sedimentation were formed by the elevated sea level: a littoral sandy zone of beach ridges and dunes; a clayey zone of tidal flats, salt marshes and brackish lagoons; and beyond this, a zone of peat deposits in a freshwater environment. These zones shifted to the east as the sea gradually flooded the formerly dry North Sea floor. In order to understand this process, one has to remember that freshwater, which is lighter than seawater, floats on the sub-surface body of salt water formed by penetration of salt water into the land subsurface during encroachment by the sea. Thus, the interface between the saline and freshwater moved eastward simultaneously with the movement of the seashore.

2.4.1.c CLIMATE CHANGES DURING THE HOLOCENE
The western area
A chronology of transgressive and regressive phases for the Netherlands' western coastal plain during the Holocene was suggested by Hageman (1969) and modified by Jelgersma *et al.* (1970) and Zagwijn and Van Staalduinen (1975) (Fig. 2.6). These studies suggest that the western part of the Netherlands is composed of layers deposited during the Holocene. A cross section consists of marine sand and clay deposits in the west and tidal flat deposits, peat and lagoonal beds in the east. The marine layers in the western coastal part of the Netherlands are overlain by a series

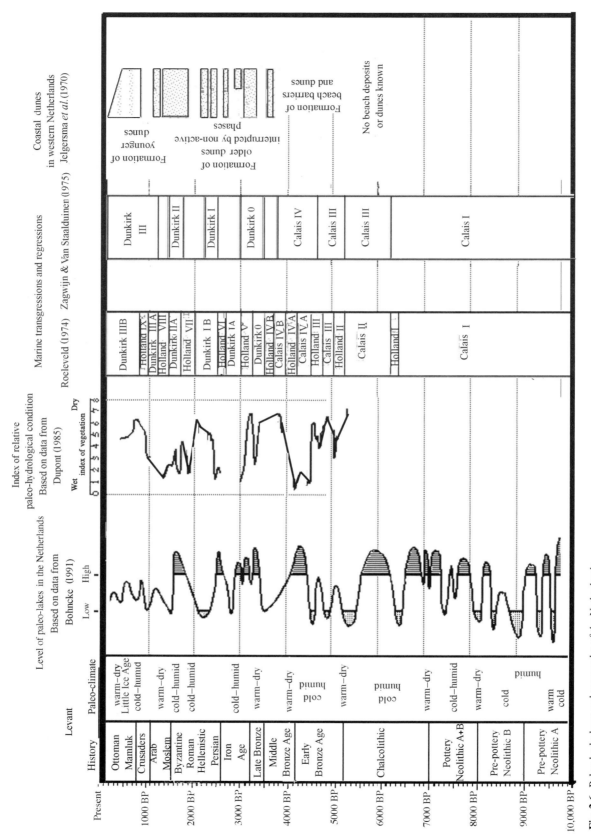

Fig. 2.6. Paleo-hydrology and geography of the Netherlands.

of old and new dunes, with paleo-soils and peat layers sandwiched between them.

Sea floor and barrier deposits of the Calais transgressions overlay the Pleistocene age sand in the western coastal part of the Netherlands. The layers of the lowest Calais I transgression are found only in the facies of the inner shore deposits of peat and lagoonal beds. The Calais II transgressions are present as tidal flat deposits, while those of Calais III and Calais IV form coastal barrier ridges. The dating of the peat layers, which were deposited simultaneously with the barrier ridge of Calais III, shows that this transgression took place c. 4.8 ka BP. Late Neolithic settlements, found on top of the eolian deposits above Calais III, show that the phase of the Calais III transgression had ended before 4.2 ka BP. Therefore, the Calais IV transgression started a little before 4.2 ka BP and ended c. 3.5 BP.

The deposits of the Calais transgressions are overlain along the present coast by a series of sand dunes, called the Old Dunes. This section of dunes is interspersed with thin soil and peat layers, showing that the dunes were deposited during a few phases of transgressions. These are correlated with the transgressional Dunkirk phases. The complex of Old Dunes is covered by younger dunes, which began to be deposited c. 1.2 ka BP. According to Jelgersma et al. (1970), the formation of the barrier ridges – as well as the dune complexes of the Netherlands – was dependent on the supply of sand. During the lower part of the Holocene, this supply mainly came from the sea floor. Therefore, every transgression led to the destruction of the former barrier ridges and caused redeposition of barrier ridges along new shorelines. A small percentage of the sediments came from rivers.

The sea level stopped rising c. 5 ka BP, at which time the coastline was located 10 km east of the present one. As the coastline shifted westward, coastal barriers and three tidal inlets were formed. The construction of coastal ridge barriers was completed c. 3 ka BP. The formation of the Old Dunes, overlying the coastal ridge barriers, ended c. 2 ka BP, namely during the Roman period. Jelgersma et al. (1970) believe this was caused by depletion of the sand. They do not consider a regressional phase at this period to be the reason for termination of the dune formation, which is the explanation I would favour. The renewal of dune deposition, the Younger Dune phase, started c. 0.8 ka BP. Again, this is not linked to a transgressional phase but to processes of shore erosion, which may be connected with the Little Ice Age. They also link the formation of the dunes to the cutting down of the forests. The processes that formed the sand dunes may have caused shore erosion through removal of the sand along the shore. Yet, Jelgersma et al., do recognize the fact that the dunes reveal cyclic variations between over blowing and soil formation. The soil formation phase is indicative of wet conditions whereas sand removal indicates dry conditions. These variations occurred simultaneously over the whole area, suggesting a common cause. However, the anthropogenic reason for formation of the Young Dunes does not hold for the Old Dunes, since humans did not inhabit all the land on which these dunes were deposited. Moreover, the cyclicity of the dune formation is similar to that observed in the alternation of transgressive and regressive phases in the coastal area. Any explanation must also cover cyclicity in the formation of peat layers in the hinterland of the dunes. Jelgersma et al. (1970) suggested a mechanism that would account for all these simultaneous phenomena, relating to changes in precipitation, which influences the frequency of high floods in the rivers and thus the flooding of the estuaries. In periods of higher precipitation, gale activity and force increase, affecting the shore environment. Another hypothesis relates to the over blowing of the sand during dry phases, perhaps leading to the blockage of some tidal inlets, resulting in regression in the hinterland.

Jelgersma et al. (1970) quote some studies in Dutch which suggested that transgressions and regressions in the Netherland coastal areas resulted from the actual breaking up of the coastal barrier complex. Vegetation also played a role in the process of dune formation. The vegetation was less dense during the dry than during the wet phases. Thus, over blowing began as the vegetation cover opened up locally as a consequence of reduced precipitation and soil drying.

In conclusion, Jelgersma et al. (1970) suggest three factors responsible for the processes of coastal dune building in the western Netherlands:

- formation of coastal barriers and the subsequent coastal development;
- climate (wind and precipitation);
- vegetation cover: a function of human as well as climatic interference.

They do not suggest eustatic transgressions and regressions as factors in the formation of either the coastal ridge barriers or the coastal sand dunes. Van der Woude (1983) investigated the Holocene paleo-environmental evolution of the peri-marine fluviatile area in the western Netherlands. In this area, fluviatile clastic beds (clay and sand deposits) alternate with peat layers. These overlie loamy river deposits and river dunes of the late Weichselian Early Holocene age. The history of the region during the Holocene can be deduced from analysis of boreholes and profiles utilizing lithology, pollen analysis and ^{14}C dates. From these data, it was found that the rise in the level of Holocene groundwater brought about the development of moist conditions in the region from c. 7.4 ka BP. After an initially slow organic lacustrine deposition of peat, and a precursory fluvial clay deposition, extensive fluvial deposition of clay and sand took place in a fluvio-lagoonal environment. At c. 6.1 ka BP, there was a fall in the absolute water level and the region became covered by swamp forests, which persisted in many places despite the continuation of the Holocene sea-level rise on a global scale. From *c.* 5.3 ka BP onwards, clay deposition took place. From *c.* 4.1 ka BP, extensive fluviatile deposition occurred, synchronous with transgressive marine phases. Around 3.8 ka BP, shallow open-water conditions persisted for several centuries. The complete covering by swamp forest first occurred in the downstream area *c.* 3.3 ka BP and reached the upstream area *c.* 2.7 ka BP. These conditions persisted up to *c.* 2 ka BP.

Evolution of the central part of the coast of the Netherlands in the beach-barrier area, during the Mid to Late Holocene was investigated by Van der Valk (1992). In his opinion, the main factors affecting this evolution were not paleo-climatic but rather the dynamics of a coastal sedimentary system, controlled, on one hand, by a gradual rise in sea level and, on the other, by the processes of progradation, related to storm wave action.

A synthesis of the studies carried out on the Holocene's paleo-environments in the western part of the Netherlands offers the following sequence of events. During the transition period, from the Weichselian Late Glacial to the Early Holocene, both fluvial deposits and the eolian river dune sands were deposited. From *c.* 7 ka to *c.* 5.6 ka BP, not much fluvial activity is in evidence. The deposits were mainly of humic clays and peat. The paleo-botanical evidence points to quiet shallow water conditions. From *c.* 5.6 ka to *c.* 5.1 ka BP, a levee-forming stream, developing a stream ridge, flowed through this area. To the south, in the basin area, aquatic conditions existed, causing a significant influx of clastic material. From *c.* 5.1 ka to 4.1 ka BP, there was a prolonged phase of reduced fluvial activity. There was a marshy environment in which peat and organic clays were deposited. During a short interval around 4.1 ka BP, a major branch of the river Rhine formed a sandy levee containing some gravels. From *c.* 4.1 ka to 3.45 ka BP, fluvial conditions were again reduced and deposition of organic material persisted in a semi-terrestrial environment. It is supposed that the area was flooded during winter and spring, while dry conditions prevailed during summer and autumn. From *c.* 3.45 ka to 3.15 ka BP, fluvial conditions prevailed and clays were deposited. These conditions may be related to the fact that a major branch of the Rhine was situated not far to the northeast. Later on, the basin area was characterized by aquatic conditions. From *c.* 3.15 ka to 2.1 ka BP, mainly shallow aquatic conditions, with abundant vegetation, prevailed. Between 2.1 ka and 1.7 ka BP, deposition of fluvial clay in the basin area occurred. From 1.7 ka to 1.6 ka BP, semi-terrestrial to very shallow aquatic conditions existed. From 1.6 ka to 1.5 ka BP, the area became wetter. Later on, significant fluvial sedimentation occurred.

The northern region
Roeleveld (1974) investigated the Groningen coastal area in the northern region of the Netherlands (Fig. 2.6). He distinguished two major types of sediment: clastic material and peat. He maintained that the clastic material was deposited as a result of tidal action while the peat was formed along the inland margins of the basins in which the clastic materials were deposited. Roeleveld argued that the classical division of the Holocene along the northwestern coast of Europe (northern France, Belgium and the Netherlands) was over simplified. This division was based on a lower clastic layer – Calais deposits – an intermediate peat layer – Holland peat – and an upper clastic layer – Dunkirk deposits. The data from the northern Netherlands and the North Sea coastal district in Germany revealed the existence of a rather complex alternation of clastic layers and peat. Moreover, it showed that a continuous but gradually slowing rise in the level of the sea took place during the Holocene, while cyclic variations, with a periodicity of *c.* 500 years, occurred at the same time. These variations were interpreted as signs of transgressions and regressions. According to Roeleveld (1974), the mechanism behind these cycles is not certain. However, the transgression and regression phases were dated and correlated with archaeological periods, as well as with other divisions of the Holocene in other parts of the Netherlands.

Roeleveld constructed a curve of the rising sea level in the Groeningen area. It was based on the relationship between the radiocarbon dates, representing phases of regressive maxima, and correlated with the curve proposed by Louwe Kooijmans (1974). He found an agreement between the two curves, thus reinforcing their validity.

The overall picture of the geological evolution of the Groeningen coastal area according to Roeleveld's curve is as follows:

1. A general sea-level rise, reflected in an overall transgressive development;
2. A decrease in the rate of sea-level rise during certain periods, especially around 3.6 ka BP;
3. The supra-regional transgressive and regressive oscillations recorded in the chronological regularity of the occurrence of transgressive and regressive phases in the area of Groningen.

Louwe Kooijmans (1980) correlated the archaeological sites with the coastal changes of the Netherlands, which were governed by the rise and retreat of sea level. During transgression phases, estuarine creek systems gradually became extended and flat tidal areas were enlarged. This was followed by periodic sedimentation, causing the creek systems to fill up with silt, and the tidal flats to change into salt marshes. This sedimentation phase was included in the transgression part of the cycle, but it represented, in essence, the first part of a regression phase, which culminated in widespread peat formation.

Casparie (1972) investigated the stratigraphy and development of the Late Glacial and Holocene peat deposits in the northeastern part of the Netherlands. His pollen analyses show that the transition zone from the Pleistocene to the Holocene, between *c.* 11.8 ka and 10.9 ka BP, was characterized by an increase in arboreal pollen

(*Juniperus*, *Salix*, *Betula* and *Pinus* spp.). From *c.* 10.9 ka to *c.* 10 ka BP, there was a marked phase of loess blowing into the region, forming mud deposits. During the same period, there was a decrease in arboreal pollen, especially *Pinus*, and an increase in Cyperaceae and other heliophilous plants. Towards 10.3 ka BP, the predominant vegetation of this loess deposit was a small birch tree. Later, a non-ferruginous fen peat was formed in depressions, which gradually extended until *c.* 7.5 ka BP. During this period, there was a drying process of the peat bog, which continued until *c.* 7 ka BP. The quantity of arboreal pollen also increased and *Corylus*, *Ulmus* and *Quercus* spp. appeared.

Shortly before 7 ka BP the area became moist again. Fen-wood peat containing *Alnus* and *Betula* developed quickly but then stopped when the supply of water decreased at *c.* 6.5 ka BP. At this time, seepages of ferruginous water developed in the eastern part of the region, again causing the formation of ferruginous peat deposits. The supply of seepage water increased *c.* 6.5 ka BP to such an extent that it flooded and covered the non-ferruginous fen-wood peat area for a few centuries. Later, *c.* 6 ka BP, highly humified peat of *Sphagna* spp. established itself on a large scale in the *Pinus* forest, and this caused the decline of the forest. By *c.* 6 ka BP, only a few trees remained from the previous uninterrupted pine forest. By *c.* 5.1 ka BP, the seepage had stopped and the peat bog had dried out. This allowed the *Pinus* woods to reestablish in areas where the iron content was not too high. Within a short time, the area became moist again, and within 150 years the highly humified sphagnum peat again overgrew the *Pinus* forest. About 4.5 ka BP, part of the sphagnum peat drained via drying cracks in the seepage peat. Then, *c.* 4 ka BP, highly humified peat growth took place. At *c.* 3.5 ka BP, an elongated lake with no outlet was formed in the region. The moisture content of the peat continuously increased until, *c.* 2.5 ka BP, the water spilled over to form a rivulet and many lakes and pools emptied, causing extreme erosion.

Dupont (1985) studied the paleo-ecology of the raised bog system in the northeastern region of the Netherlands (Meerstalblok; Fig. 2.6). The study involved the analysis of pollen assemblages, the determining of pollen density as a function of time, the identification of macro wood remains, as well as the determination of deuterium to hydrogen isotopes ratios in the cellulose of the peat. The systematic ^{14}C dating of the research profiles enabled the authors to calibrate the ^{14}C dates with the dendrochronological time scale suggested by Klein *et al.* (1982). The stratigraphy was established on the basis of two sections. The comparison between them showed that layers of the same age lay about 40 cm lower in one section than in the other (distance between sections about 3 m). They concluded that one was deposited on a hummock while the other was in a hollow. The last section showed a hiatus in the sedimentation, which started *c.* 2.5 ka BP (dendrochronological age, *c.* 2.4 ka BP on the radiocarbon time scale (^{14}C BP))

and lasted for about 400 years. Dupont explained that this was a result of an increase in rainfall to a level above the bog's retention capacity, causing a drainage rivulet to be formed. The influence of humans can be traced from *c.* 5.5 ka BP dendrochronological years (*c.* 4.8 ka years BP: the start of the cultures named Funnel Beaker and Protruding Foot Beaker).

The analysis of all the components mentioned above allowed a temperature and humidity curve for the Holocene to be constructed. The main conclusion to be drawn from this curve is that cold periods equate to more humidity because of an increase in precipitation. For example, this was the reason for the hiatus of deposition at 2.5 ka BP dendrochronological years (*c.* 2.4 ka ^{14}C BP).

Detailed work on the paleo-hydrological changes in the Netherlands during the last 13 ka was carried out by investigating the changes in both fluvial and mire environments, as expressed by changes in the paleo-vegetation assemblages (Bohncke and Vanderberghe, 1991). It appears that the beginning of the last deglaciation, following the Weichselian pleniglacial period, was characterized by a superfluous supply of eolian material owing to lack of vegetation cover and low precipitation. At *c.* 13 ka BP, as temperatures and precipitation rose, the bare soil became covered by tundra type grasses and herbs, thus reducing erosion and enabling the soils to stabilize. From *c.* 12 ka BP, evapo-transpiration increased, which caused a fall in the groundwater table. This, in turn, resulted in local dune formation. A severe cold period occurred between *c.* 10.85 ka and 10.5 ka BP, while the period between 10.5 ka and 10.25 ka was dry. The establishment of a more oceanic humid climate *c.* 8.3 ka BP led to a general rise in the groundwater table. After the climatic optimum (*c.* 8.5 ka to 6 ka BP), the effective precipitation increased.

The paleo-environment and paleo-hydrology of the Holocene in the Netherlands was derived from a detailed study of the paleo-flora in the sediments of Mekelermeer, a small lake in the northern part of the country. During its history, the lake went through a sequence of changes in its level, which, in the opinion of Bohncke (1991), was a function of the changes in humidity during these periods. During wet periods (with abundant precipitation), the level of the lake and that of the groundwater in its vicinity rose, while during dry periods they fell. These fluctuations are represented in Fig. 2.6.

Bohncke (1991) did not investigate the correlation between humidity and temperatures. Yet, from several correlations, one can deduce that wet periods during the Holocene were coincident with cold periods, while dry periods were the result of a warm climate. For example, the decline in the abundance of *Corylus* started *c.* 2.6 ka BP, during a period of a high water table. This decline in *Corylus* is a well-known feature in many northwestern European pollen profiles and is generally interpreted as being caused by the climate getting colder and wetter. (See references in Bohncke,

1991.) A correspondence also exists between the period of high water table c. 0.7 ka BP and low temperatures recorded in England during that period (Aaby, 1976). Moreover, a predominant positive relationship was found between the start of transgression phases and phases with low lake levels, recorded in the Noordoost polder region in the northern part of the Netherlands by Van de Plassche (1982). In this study, a time relative mean sea-level curve for the period c. 7.5 ka to 2 ka BP was constructed. This curve was obtained from the re-evaluation of earlier curves suggested by various authors, as well as from new data derived from investigating past fluctuations in sea level revealed by studying ancient beach plain sediments bordering the former Old Rhine estuary. The groundwater level curve was derived from basal peat data and the paleo-altitude of river dunes (donken). Van de Plassche (1982) found that the beach plain curve was more or less parallel with the revised mean sea-level curve: the curve for the river dunes lying slightly above the other until c. 6.7 ka BP and later converging with it. According to Van de Plassche, this is because, until this date, the morphological features of the area were affected mainly by the river gradient, while later the flood plain influence became dominant. In accordance with Van der Woude (1983), this author found a positive correlation between the sea-level fluctuations and the levels of the peat deposits, which are dependent on the level of the groundwater. He also found that the fluctuations which he observed in the sea-level curve corresponded to the transgressive and regressive intervals.

Steenbeek (1990) investigated the stratigraphy of two sites in the delta region of the rivers Rhine and Meuse. Deposition started from 7 ka BP in the more western sites and from 6.1 ka BP in the more eastern site, at a distance of about 23 km to the east.

The onset of vertical accretion thus reflected the adjustment of the profile along the river to the rising sea level, causing a progressive shift inland, as well as an intensified rate of deposition. The latter was not merely a function of increased river discharge but was primarily caused by a rise in sea level. The levels of the levee and basin, simultaneously deposited, show an altitudinal difference of 1.5 m. Thus, a general picture emerges of fluvial deposition during the Holocene in this region governed mainly by the rise in sea level, progressing upstream. Fluvial clastic sediments were deposited in the area in the form of fluvial clays and gyttjas (organic clays), while reduced fluvial activity allowed peat formation. The rate of deposition of the fluvial material in the peri-marine areas could not counterbalance the process of accretion brought about by the rise in the regional water level. This caused the formation of large areas in which fluvio-lagoonal and fluvio-lacustrine conditions prevailed. However, in more terrestrial zones, the gradients remained much steeper and differences between channel and basin levels persisted.

The factors that could have caused the various changes in the physical and human environment observed during the period 1000 to 1300 AD were discussed and analyzed in a special symposium held at the University of Amsterdam in 1983 (Berendsen and Zagwijn, 1984). The main changes observed for this period were as follows. Increased river recharge occurred, especially from 850 to 1000 AD, when increased precipitation was reported, and from 1250 to 1400 AD. A relative increase in river flooding was observed for periods during the ninth, eleventh, thirteenth, fifteenth and sixteenth centuries AD. However, the tenth century was relatively dry. The number of severe storm surges seems to have increased a little after 1200 AD and reached a maximum in the sixteenth century. In the twelfth and thirteenth centuries, many storm surges in the northern and southern parts of the Netherlands led to the widening of the tidal inlets and to the enlargement of the estuaries. Thus, tidal influences reached further upstream. A number of authors believe that a sea-level transgression occurred between 800 and 1000 AD. Some of these authors believe that this transgression was related to the formation of the Younger Dunes, which received their material from the submarine erosion of the coastal barrier. These phenomena were related to coastal erosion and the coastal profile becoming steeper. In the dune area, peat growth ended between 900 and 1000 AD. Under relatively dry hydrologic conditions, the dune belt widened considerably from about 1000 to 1180 AD. Then, from 1180 to 1330 AD, the water table in the dune area rose and there was an increase in vegetation. These conditions continued until 1600 AD.

In northern Netherlands, clays were deposited under brackish conditions between 1150 and 1250 AD. The population of the area increased between 1100 and 1300 AD. During this time, forests were cleared, peat was extracted for fuel and salt, and the following three rivers were dammed: the Kromme Rhine (1122 AD), the Ijssel (1285 AD) and the Linge (1304 AD).

Van Geel and Renssen (1998) suggested that a cold period occurred between 850 and 760 calendar years BC (2.75 ka to 2.45 ka ^{14}C BP) because of reduced solar activity.

Conclusions

Regarding the relationships between the fluctuations of sea level and groundwater table, one can see that there is a difference between the peri-marine environments described by Van der Woude (1983) and Van de Plassche (1982) and the inland areas described by Bohncke (1991). In the peri-marine environments, the warm climate (which correlates with high sea levels) caused a rise in groundwater level and the formation of fresh-water peat layers. At the same time, further inland, the warm climate caused lowering of the groundwater table and drying of the peat bogs.

As most of the proxy-data time series of the Netherlands are concerned with local changes, where local circumstances may have played an important role in deciding the variations, correlation with the sequence of the paleo-climates of the Levant is rather difficult. Moreover, the moderating influence of the sea may have blurred the impact of minor climate changes. However, when the major changes are compared (Fig. 2.6), such as, for example, the warm period of the MB of the Levant (from *c.* 4 ka BP to *c.* 3.5 BP), the Netherlands experienced a low groundwater table and low lake levels, while during most of the cold EB period in the Levant (5 ka to 4 ka BP), the water levels in the Netherlands were high. During the Moslem–Arab warm period from *c.* 1200 to *c.* 1000 BP (800 to 1000 AD), water levels in the Netherlands were again low. During this period, a transgression occurred that caused the invasion of young dunes. During the Little Ice Age, there was a rise in the groundwater table. The regressive conditions during this period stopped the supply of sand to the coastal dunes, which caused their fixation by vegetation.

2.4.2 The British Isles

According to Roberts (1989), the British Isles remained largely treeless until the start of the Holocene. Later on, once the climatic conditions permitted, the dominant vegetation was mixed deciduous forest. During the Lower Holocene, the Neolithic, the climate was rather warm, enabling the expansion of the hazel population. During the Chalcolithic and EB, there was a reduction in this population, most probably because of a colder phase. During the MB and Late Bronze Age hazel increased in numbers, while towards the Roman period there was a strong decrease, indicating colder conditions. According to Lamb (1982), the Chalcolithic period, from *c.* 6 ka to 5.2 ka BP, was a world-wide colder episode: most probably, a mini-glacial period. It was followed by a warm period beginning *c.* 5 ka BP. This period is called the Sub-Boreal in Europe. Although phases as warm as any since the Last Ice Age occurred during this period, the variability of the climate was greater than before. There followed another cold period, extending from *c.* 5 ka to 4.2 ka BP.

Paleo-ecological investigations in northern Scotland, using radiocarbon dating, dendrochronology and fine temporal-resolution palynological data (Gear and Huntley, 1991), show that pine forests expanded northward by up to 80 km after *c.* 4.4 ka BP. At *c.* 3.8 ka BP, the forests retreated to their earlier limits, which were similar to those of the present. Gear and Huntley (1991) explain these findings as the result of a phase of a warmer climate, which caused the drying of the blanket of mire surfaces. The warm climate was a function of the movement northward of the Azores high in summer and a consequent shift northward of the jet stream. This reduced rainfall in northern Britain and increased summer temperatures.

Bohncke (1988) carried out a palynological analysis of the peat sections on the Isle of Lewis, Outer Hebrides. The sampling site was close to the site of prehistoric megalithic monuments. Three major clearance and tillage zones were found at *c.* 5000 BP, *c.* 4200 BP and from *c.* 2520 to 2030 BP. The two first periods correspond to warm periods in the Levant, while only the beginning of the third period was a warm one (see Fig. 6.1, p. 106).

Analysis of blanket peat at a site in southern Scotland (Chambers *et al.*, 1997) provided evidence for particular wet episodes at *c.* 3.45 ka, 2.6 ka, 1.9 ka and 1 ka BP and other wet episodes at *c.* 3 ka, 2.2 ka and 1.7 ka BP. The last prolonged and particular wet (or cool and wet) episode commenced at *c.* 0.54 ka BP (cal. 1410 AD) and may be correlated with the Little Ice Age. Comparing these observations with those of Gear and Huntley (1991), one may conclude that wet periods were also periods of cold climate. Additional evidence for an abrupt change to a wetter climate between 3.9 ka and 3.5 ka BP was found on the basis of paleo-hydrological and paleo-ecological investigations in northern Scotland (Anderson et al. 1998).

It can be inferred that the sea level was lower during the Roman period, and the climate was presumably colder, from the fact that many Roman buildings built prior to 300 AD were covered by transgressions of the sea that occurred shortly thereafter (Thompson, 1980). This rise in sea levels most probably indicates a short warming-up phase around 300 AD. According to Lamb (1982, p. 86), the whole period from 300 to 1150 AD was warm, except for a downturn from 400 to 500 AD.

Brown (1998), on the basis of various sources of evidence, tended to the opinion that the climate of the British Isles was rather cold during the Roman period and was being influenced by a high frequency of blocking anticyclones over Scandinavia, especially during winter. This caused a lower frequency of southwesterly storms. The groundwater level was high during this period.

Lamb (1977, 1984a) reconstructed the temperatures for central England from 800 AD to the present and the variations in sea surface temperatures of the northeast Atlantic. He maintained that over northern and central Europe, during the warmest periods, the long-term average quantities of precipitation were between 10 and 20% higher than now, while during the Little Ice Age, they were less than the present by approximately 5 to 15%.

Palmer (1986) used a general atmospheric circulation model to test situations in which abnormally cold sea surface water has been claimed by researchers for the sea near the coast of Newfoundland. Such situations may be connected with a weaker Gulf Stream. He found similarities between observations and the output from a climatological model in which cold sea surface anomalies correspond to high-pressure anomalies over Europe. When the sea surface temperatures were warm, the westerly type of weather dominated, which caused higher rates of precipitation over Great Britain.

2.4.3 Scandinavia and the northern Atlantic

2.4.3.a SCANDINAVIA

After the retreat of the glaciers of the last glacial period, which covered Scandinavia, a Baltic ice lake was formed *c.* 13 ka BP. It existed with some interruptions until *c.* 10.3 ka BP. It was replaced by the Yoldia Sea, which existed until 9.5 ka BP. At *c.* 9.9 ka BP, a saline ingression increased the salinity of the water of this sea. Then, from 9.5 ka to 8 ka BP, the area was covered by a vast freshwater lake (Ancylus Lake), which was created by the retreat of the sea. Finally, from 8 ka BP to the present, the brackish inland sea conditions characteristic of the contemporary Baltic Sea have existed (Björk, 1995).

The first stratigraphical section for the post-glacial periods was established in Scandinavia by the Swedish and Norwegian botanists Rutger Sernander and Axel Blytt (Blytt, 1876; Sernander 1894) on the basis of the pioneering work by the Danish botanists

Heinrich Dau, Japetus Steenstrup and Christian Vaupell, who investigated the flora of the bogs in their country. (For a fuller description of the history of the paleo-botanical research in Denmark as well as the detailed description of Denmark's environmental history, on which the following section is based, see Iversen (1973) and Bradley (1999, p. 14).) The Blytt–Sernander stratigraphical column distinguishes four periods evidenced by changes of the flora of the bogs and caused by severe climate changes. The periods are the Boreal (dry), the Atlantic (humid), the Sub-Boreal (dry and warm) and the Sub-Atlantic (humid and cool) (Fig. 2.7). As

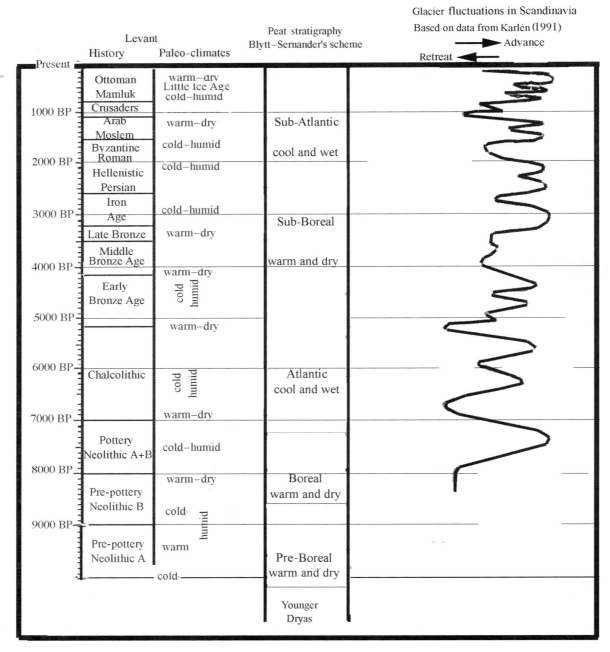

Fig. 2.7. Holocene geology of Scandinavia.

palynological research and ^{14}C dating methods were developed, further divisions of the stratigraphical column were correlated with the Blytt–Sernander division (Bradley, 1999). A preliminary correlation between this division and that of the Levant (Fig. 2.7) shows clearly that only the long and prominent climate regimes were marked. This is true for the Boreal, which is equivalent to the warm part of the Neolithic; the Atlantic, which is equivalent to the cold Chalcolithic and possibly the EB; the warm Sub-Boreal, which is equivalent to the MB and Late Bronze Age; and the Sub-Atlantic, which covers the Roman–Byzantine, Crusader and Little Ice Age cold periods. Consequently, although the Blytt–Sernander division is important from a scientific historical point of view, to-day it is outdated and it is regrettable that it is still being referred to by contemporary researchers.

The boundary between the last glacial period and last interglacial period is determined in Scandinavia according to the retreat of the ice sheet from its southernmost point in mid-Jutland. However, the boundary in pollen diagrams is set with the first appearance of *Artemisia*, as this is regarded as the first clear indication of a warmer climate. This was found in a section from Lake Bölling to have appeared *c.* 12.5 BP. Before that, the fauna was that of tundra, characterized by *Dryas (D. aven)* and dwarf willow; consequently this period is known as the Oldest Dryas. The first tangible warm phase started *c.* 12.5 ka BP and continued to 12 ka BP. It was characterized by the appearance of *Betula* (birch). This period is called the Bølling after the place where it was first identified. It seems that the vegetation in Denmark at that time was that of "park tundra", with birch woods in the warm places and tundra in the cold north-facing or damp areas. Moreover, vast stretches of the country were still covered by remnants of the ice cap or "dead ice". From about 12 ka to 11.7 ka BP, a cold short phase followed, the Older Dryas. During this phase, the trees disappeared and tundra conditions became dominant. About 11.7 ka BP, birch and other forest trees reappeared, which indicates a warming up of the climate. This is called the Allerød period, based on the locality in north Zealand where it was first recognized. In this place, a layer of black organic deposit (gyttja) was observed, containing remnants of forest flora and fauna. The forest was continuous in the southeastern part of Denmark, where July temperatures reached an average of 13–14 °C.

About 11 ka BP, a sudden decrease in temperature occurred, an event called the Younger Dryas, which continued until 10.8 ka BP. Again park tundra replaced the forest in the southwest of Denmark, while full tundra conditions prevailed in the northwest. The fall in the average temperature from the preceding period was approximately 3–4 °C. Precipitation during the cold periods was lower than that during the warm periods but soil moisture was higher, because evaporation rates were low. Snowfall during the warmer periods seems to have been moderate, based on the abundance of pollen of flora that cannot tolerate prolonged cover by snow.

Deep-sea sediment cores from the Atlantic show that the polar front shifted as far north as Iceland during the interstadial of 13 ka to 11 ka BP (Roberts, 1989, p. 51–52; Ruddiman and McIntyre, 1981).

The Holocene started *c.* 10.3 ka BP with a steep rise in temperatures, changing the open landscape of the Younger Dryas into a continuous forest. It began with the rapid spread of juniper scrub, followed by birch, aspen and pine. Evaporation as a consequence of the warm climate and transpiration by the forests caused lowering of the groundwater table, and many shallow lakes became overgrown. At *c.* 10 ka BP, there occurred a short recession in the warm climatic conditions: the Friesland oscillation. After this short cold period, the climate warmed up again until *c.* 9 ka BP. This warm period is called the Pre-Boreal in Blytt–Sernander's division and was renamed the Birch–Pine period by Iversen (1973).

At *c.* 9 ka BP, the climate became even warmer and was characterized by forest vegetation dominated by hazel (*Corylus*) and pine. This is the Boreal period of Blytt–Sernander and the Hazel–Pine period of Iversen. This type of vegetation continued up to *c.* 8 ka BP, when the forest vegetation became more varied. Among the trees to appear were the lime (*Tilia*), the oak (*Quercus*) and the alder (*Alnus*). This is the Atlantic period of Blytt–Sernander and the Older Lime period of Iversen. It lasted until 5 ka BP. Thereafter, the pollen assemblage started to show the introduction of domesticated plants, as farming communities settled in the region. According to Iversen (1973), the later half of the Older Lime period is believed to be the warmest of the post-glacial periods, with temperatures at least 20 °C higher than today.

The period from 5 ka to 2.5 ka BP – termed by Iversen the Younger Lime period and correlated with the Sub-Boreal of Blytt–Sernander was still dominated by the lime forest, but a decline in elm and ivy occurred. The climate during this period was assumed by Iversen to have been warm, but the first signs of a decline in temperature are in evidence; however, he does not maintain that the pronounced reduction of the elm was a function of climatic changes – although this was the reason that Blytt and Sernander set the border between the Atlantic and Sub-Boreal divisions at this time. According to Iversen (1973), this disappearance is still a mystery and he attributed it either to Dutch elm disease or, more probably, to the influence of the primitive peasant culture with its domesticated animals.

According to Sernander (1894), the passage from the Sub-Boreal to the Sub-Atlantic was characterized by deterioration in the climate. This happened *c.* 2.5 ka BP. Iversen (1973) refers to this period as the Beech period. He maintains that this was also the period in which human interference in the ecological balance of the forest became very pronounced, especially in forest clearances.

In addition to stratigraphy based on pollen sections, Iversen also mentioned the raised bogs as climatic indicators. These formed huge sponge-like carpets, composed mainly of the peat sphagnum moss. As this moss is highly dependent on rainwater, which it absorbs through pores in its leaves and stores there, it flourishes during cool and humid periods. Consequently, profiles that show changes from dark peat, rich in heather vegetation, to a lighter one, rich in sphagnum, are indications of climate changes from drier to more humid conditions. Such a boundary sets the dividing line between the Sub-Boreal and the Sub-Atlantic and is also the dividing line between the north European Bronze and Iron Ages.

According to Iversen (1973), the Swedish scientist Granlund observed five such boundary layers, Gh I to GH V, occurring approximately at 0.8 ka, 1.6 ka, 2.6 ka, 3.2 ka and 4.3 ka BP, respectively. Of these Gh III was the most widespread. According to Iversen (1973), the Sub-Atlantic deterioration in the climate is also seen in the advance of the glaciers in Alaska, which corresponds with Gh I, II and III.

Aaby (1976) investigated a number of large open peat sections in five raised bogs in Denmark. Although he accepts that past climatic changes are reflected in raised bogs as variations in the degree of decomposition or humification of the peat, he maintains that interpretations are not straightforward. For example, the change into lighter-colored bog, which shows less humification, may be a result either of colder climate or more precipitation, or both. Nevertheless, his general conclusion is that light-colored peat was formed when wet periods were more frequent than dry ones, and vice versa. In addition to the degree of humification, he also investigated the relative distribution of two rhizopod genera, which are also indicators of the water regime of the bogs. He concluded that the bogs indicate long-term cyclic climatic variations, with a periodicity of about 260 years. He recommend that these results should be used to model future climates.

As was already discussed, the Blytt–Sernander climatic divisions determine the stratigraphy suggested by Iversen (1973), although a more critical approach could have shown the limitations of the former to describe the more detailed picture. For example, an examination of the pollen diagrams from Lake Bølling, as well as those from eastern Denmark, presented by Iversen in his book of 1973 clearly shows additional variations on top of those based on the Blytt–Sernander division.

These variations can be clearly seen in the curve of glacier fluctuations in Scandinavia during the last 9000 years (Karlén, 1991, p. 409; Fig. 2.7). The data for the curve were derived from studies of sediments from lakes downstream of small glaciers and radiocarbon dating of organic material found in the sediments. Karlén (1991) found that the mass balance of the glaciers depends on the summer temperatures in Sweden while along the Norwegian coast, it depends on winter precipitation. A correlation was also found between the mass balance of the glaciers and the tree ring records from Sweden. Narrow tree rings correlated with a positive balance and vice versa. Karlén and Kuylenstierna (1996) correlated climate change in Scandinavia, as evidenced by the advance

of glaciers and fluctuations in the tree line, with the changes in solar irradiation as evidenced by $\delta^{14}C$ anomalies. They found that 17 of 19 events of low solar activity coincide with periods of cold climates.

In central southern Norway, a period of glacier advance took place between 9 ka and 8 ka BP. Another glacial advance was found to have occurred c. 7.5 ka BP (^{14}C date) and another around 1 ka BP (Nesje and Dahl, 1991). Dahl and Nesje (1996) developed a new way to calculate winter precipitation during the Holocene by combining glacier equilibrium-line altitudes and the occurrence of pine trees. They found that the wettest phase was c. 8500–8300 BP. During this period, summer temperature was approximately 1.35 °C warmer than the present. These conditions changed abruptly, within 30–50 years, to a regime dominated by dry winters and summers that were a little warmer than the present. The transition can be correlated with an abrupt change into lighter oxygen isotopes recorded in Greenland ice cores.

Lamb (1984a, p. 234) cites Holmsen, who described the area covered by farming in central Norway. This was more or less unchanged from the Early Iron Age but retreated in places after 400 AD and spread abruptly between 800 and 1000 AD.

Mörner and Wallin (1977) analyzed oxygen and carbon isotopes in the carbonate sediments of a lake on the island of Gotland in the Baltic Sea. They converted the isotopes ratios into temperature based on the relation between measured lake temperatures and corresponding isotope composition of the water. Their temperature versus age curve showed minimum temperatures between 10 ka and c. 9.3 ka BP. Later, until c. 8.5 ka BP, came a period of average temperatures a little above the present. This was followed by a warmer period extending for a few centuries and returning to average at c. 8.2 ka BP. Then came a warmer period to c. 5.8 ka BP. From 5.8 ka to 2.5 ka BP, there was a period of temperatures fluctuating a little above and around the present average. At c. 2.5 ka BP, it became colder until c. 1.2 ka BP, when there was a slight change. At the end of the section, at c. 1 ka BP, a colder trend can be seen.

Mörner (1978–79), summarizing data on sea level changes along the northwestern European coasts, concluded that eustatic fluctuations could be discerned in addition to the isostatic rise of Fennoscandia (830 m since the Late Weichselian glaciation). He maintained that the eustatic fluctuations correlate with paleomagnetism and paleo-temperatures, which suggests, in his opinion, a mutual origin. He identified and dated some 40 shorelines in the Kategat region. Because of tilting of the Fennoscandian block, the effect of the eustatic transgressions decreases inland, whereas the effect of the regressions increases inland. His curves are given in sidereal years, corrected against ^{14}C dates, on the basis of dendrochronology and Swedish varve chronology. He observes a "regression maximum" from 9.75 ka to 9.3 ka BP, a distinct marine transgression from 9.3 ka to 8.3 ka BP, another distinct regression from 8.3 ka to 8 ka BP and after 8 ka BP, a transgression which reached its maximum at 7.7 ka sidereal years BP. The later oscillations were relatively small.

Ambrosiani (1984), summarizing his investigations on sea levels in Sweden, showed that the almost linear isostatic up-heaval during the Post Glacial Period, causing a rise of land height of about 0.5 m per century, slowed down to 0.35 and even 0.25 m during certain periods. Consequently, the retreating shoreline curve assumes a step-like, rather than a smooth, curve. He attributed these changes to climate changes, that is, eustatic changes.

Harrison *et al.* (1993) used data of changing levels in seven lakes in southern Sweden, and data from Estonia, France and Greece, in a water balance model to quantify the effects of evaporation, as a function of insolation, temperatures and cloudiness, on runoff. The data from southern Sweden showed high water levels at 10 ka BP; a fall from *c.* 10 ka to 9.5 ka BP, with a minimum reached *c.* 9 ka BP; and higher lake levels returning between 8.5 and 6 ka BP. Most lake levels were low *c.* 4 ka BP. Later, there was a general rise, with a relatively low level between 1.5 ka and 1 ka BP.

By analyzing bog formation in southern Sweden, Svensson (1988) also linked cold climate with more precipitation, which caused higher water tables. He observed a change to bog veg-etation *c.* 7 ka BP, coinciding with a rise in the level of many lakes. This continued for about 1000 years and was followed by a period of drying up and humification of the peat layers and at *c.* 6 ka BP, a dry period evidenced by low lake levels was observed. Humid conditions started again at *c.* 5 ka BP. During the "Middle Sub-Boreal chronozone", presumably *c.* 4 ka BP, there was a stage of low water levels. About 2.4 ka BP, there was another stage of increased humidity. A stage of humification and low lake levels, as a function of a drier climate, was observed *c.* 1.7 ka BP. An-other period of low lake levels occurred *c.* 1.3 ka BP. About 1 ka to 1.2 ka BP, the climate became more humid.

Digerfeldt (1988) presented a similar, though more detailed, section of lake level changes in southern Sweden. His section was based primarily on sediments and their pollen assemblages in Lake Bysjö as well as in other lakes. He observed a distinct low level at *c.* 9.5 ka to 9.2 ka BP. This was followed by a humid period *c.* 7 ka BP, which corresponded to that observed by Svensson (1988). A dry period followed from *c.* 6.8 ka to 6.5 ka BP, and a colder and humid period at *c.* 5 ka BP corresponded with that observed by Svensson. According to Digerfeldt, a drier climate ensued, reaching a maximum from 4.9 ka to 4.6 ka BP. This event should be correlated with the dry event observed by Svensson at *c.* 4 ka BP. This major period of dryness continued until *c.* 2.9 ka to 2.6 ka BP. A small peak of wet conditions occurred *c.* 3.5 ka BP. The other markedly wet period was *c.* 2.5 BP. The fol-lowing change, which is also the last one observed by Digerfeldt and which peaked *c.* 1.5 BP, was dry.

Correlating the paleo-climates of the Levant with the advance and retreat curve of the glaciers of Scandinavia (Karlén, 1991; Fig. 2.8), one can see that during the main cold periods, as for example those of the Chalcolithic period (*c.* 6.5 ka to *c.* 5 ka BP), EB (*c.* 5 ka to *c.* 4 ka BP), Roman period (*c.* 2.3 to *c.* 1.7 ka BP) and the Crusader period and Little Ice Age (*c.* 1 ka to 0.4 ka BP), the Scandinavian glaciers advanced, while during the warm periods like those of the MB (4 ka to 3.5 ka BP) and Moslem–Ottoman period (0.4 ka to 0.1 ka BP), the glaciers retreated. The same cor-relation can be drawn for temperatures, precipitation rates and levels of the lakes in Scandinavia, namely that cold humid peri-ods in the Levant corresponded in general with low temperatures, high precipitation and high lake levels in Scandinavia.

Changes in microbiological environments in a lake situated in western Finnish Lapland were linked with changes in humidity and suggested a low phase from 8 ka to 4 ka BP (Hyvärinen and Albonen, 1994). As this contrasts with data from southern Finland and Sweden the authors suggest that the climate regimes in north-ern and southern Fennoscandia were different during the lower part of the Holocene.

2.4.3.b THE NORTHERN ATLANTIC

Kellogg (1984) investigated the percentage of sub-polar plank-tonic foraminifers in cores from the Denmark Strait, between Greenland and Iceland. He found high-frequency fluctuations (pe-riods of 615–784 years) in two principal species of *Globigerina*. He attributed these changes to variable dissolution rates and to changes in the boundary between the Imringer and East Greenland currents, which are a function of climate changes.

Lamb (1984b) charted the changes in the penetration of polar water southwards from East Greenland into the Atlantic. This influences the surface temperature in the North Atlantic. The sea around the Faeroe Islands, which at present is under the influence of the warm saline Gulf Stream (the average temperature of which is 7.7 °C in this region), seems to have been about 5 °C colder than the average of the twentieth century, during the climax of the Little Ice Age between 1674 and 1704. Lamb (1982) noted that Western Europe encountered the warmest temperatures of the post-glacial phase between 3.1 ka and 2.8 ka BP (1100–800 BC).

2.4.3.c GREENLAND

Although isotopic analyses of Greenland's ice cores have revolu-tionized paleo-climatic research, when it comes to the Holocene, one should take into consideration a few important constraints. In the first place, the record from Greenland is that of an area well inside the Arctic belt and, therefore, the effect of minor changes may be blurred. Indeed the isotopic records show that the Holocene was a period of relative stability, with small fluctu-ations of $\delta^{18}O$ on the order of ± 1–$2\permil$. Moreover, there is little correlation between sites, probably because of local differences in

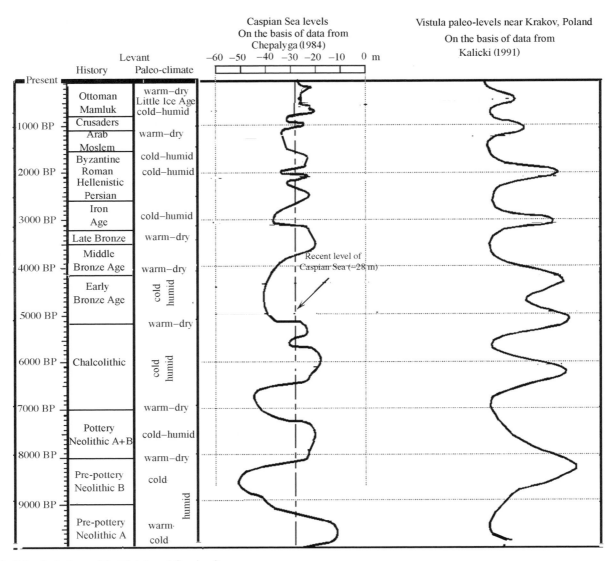

Fig. 2.8. Paleo-hydrology of the Vistula and Caspian Sea.

accumulation and wind drifting of snow (Bradley, 1999, p. 159) One would also expect that colder periods would be more pronounced in the Arctic region than in the more southern latitudes, the climate of which may be moderated by the tropical belt. A time lag between the start and end of climatic changes in both regions should also be expected. Indeed, a comparison between the Levant curves and Camp Century curve (Dansgaard *et al.,* 1971) shows such discrepancies. Changes at the beginning of the EB, the impact of which were felt *c.* 5 ka BP in the Levant, were observed in Greenland *c.* 5.3 ka BP, if not a little earlier. The same can be said for the impact of MB changes, which started in the Levant *c.* 4 ka BP but in Greenland were observed *c.* 3.5 ka BP. Also, the Roman cold period, which starts in the Levant *c.* 2.2 ka BP, its peak being *c.* 2 ka BP, reaches its peak

in Greenland *c.* 1.5 ka BP. There is, however, a rather good correlation between the two regions for the climate changes after the Roman period.

Porter (1981, 1986) demonstrated a close relation between the pattern of northern hemisphere glacier variations during the last millennium and volcanic aerosol production, as found in Greenland ice cores. He, therefore, maintained that sulfur-rich aerosols, generated by volcanic activity, are a primary factor in forcing climate change on the decade level. Glaciation lags behind the increase in acidity by about 10 to 15 years.

Jennings and Weiner (1996) have analyzed the lithofacies and the benthos foraminifer assemblages from two cores taken from Nansen Fjord, in eastern Greenland. They found evidence for the Medieval warm period between *c.* 730 and 1100 AD, an early cold interval *c.* 1370 AD and the severe cold period of the Little Ice Age from *c.* 1630 to 1900 AD. It is interesting that their record is similar to the 1000 year-long sea ice index of Iceland and, to a lesser extent, to the Crête ice core from Central Greenland.

2.5 CENTRAL AND EASTERN EUROPE

Various estimations and calculations for central Europe (more precisely for southern Poland), from the maximum extent of the last ice sheet to the present time, have made it possible to construct curves showing variations of precipitation, evaporation and runoff. Annual precipitation rates fluctuated from c. 250 mm or less at 18 ka to 14 ka BP, to c. 550–650 mm from 8.5 ka BP onwards. As stressed by Lamb (1977), annual variations were probably much higher during the unstable cooler phases.

Chernavaskaya (1990) investigated the climate of eastern Europe during the historical past by analyzing palynological assemblages of high bogs in the forest zone. From these assemblages, temperature conditions in the summer in the southwestern part of eastern Europe over the last 3500 years were reconstructed. A cooler period occurred from 3.4 ka to 1.5 ka BP. The lowest temperatures occurred during the first to fifth centuries AD. Northwards, in the Ukraine and Polessia, there was a fall in temperature in the first century AD. The last 1500 years have been a relatively warmer period. However, a colder period was observed from 900 years ago to 150 years ago, which coincided with the Little Ice Age, during which there were periods of a century in length when the amount of precipitation was 10–15% less than at present. Such periods were observed in the northern part of the region, with a time shift from the northeast (during the fourteenth and fifteenth centuries AD) to the southeast (sixteenth–seventeenth centuries AD) and the southwest (seventeenth–eighteenth centuries AD). During the Medieval Optimum (thirteenth century AD) dry periods were more typical of the southern regions.

2.5.1 Poland

Starkel (1984) analyzed the reflection of abrupt climatic changes in the relief and sequence of continental deposits in central Europe. He identified a stage of rapid cooling c. 10.8 ka to 10.6 ka BP. This caused a rise in the erosional activity of the Vistula river. Cooling and decrease in precipitation caused sand to blow from the braided river channel and dunes to form. At the beginning of the Holocene, c. 10.3 ka to 10 ka BP, he identified a warming phase, which corresponded with a decrease in fluvial activity in southern Poland, expressed in the change from channel bars to swampy over-bank deposits. At c. 8.5 ka–8 ka BP, there was a cool phase, indicated by an increase in flood frequency. In the Carpathian region, the oxbow lake mires were rapidly covered by sandy bars or by thick alluvial fans. There were also three subsequent cooler episodes: c. 5 ka to 4.5 ka BP, 2.8 ka to 2.4 ka BP and in the seventeenth to

nineteenth centuries AD, corresponding with the Little Ice Age. During all three phases, there was a rise in lake levels.

The climatic warming that followed the Little Ice Age led to the formation of dense vegetation cover and heavy summer rains. This reflected a general shift from cold winters and snow-melt runoff to summer floods. The changing frequency of extreme events when there was dense vegetation cover and, later, under increasing deforestation caused by human activity gave rise to a complicated pattern of phases with overloaded and underloaded rivers (Starkel, 1991). According to Starkel (1998), the frequency of heavy rainfalls and floods across part of the north European plain, though fluctuating, shows a strong tendency to cluster, as a reflection of changes in the atmospheric circulation. For the sedimentary and geomorphic features dating from the Little Ice Age, he finds synchronicity throughout Europe from just before 1600 AD to about 1700 AD, and then between 1800 and 1820 AD.

Kalicki (1991) also came to the conclusion that all Holocene phases of increased river activity in the Vistula valley, and other valleys of central Europe, were the result of climatic fluctuations. He found that cooler and wetter periods correlated with floods of higher frequency and magnitude and that there was also a significant increase in erosion and deposition rates in such periods (Fig. 2.8).

2.5.2 The Caspian Sea

The Caspian Sea is the largest inland lake on the globe, being c. 400,000 km^2 in extent. The catchment area covers 3.6 million km^2. The main contribution to its water comes from the Volga, which accounts for 80–85% of the total volume of annual inflow. It can, therefore, be assumed that the discharge from the Volga catchment is the main element in the hydrological balance of the lake, which has existed since the Middle Pliocene. The present level of the lake is about 28 m below MSL, while the average total salinity of the water is 12 mg/l (Mamedov, 1997).

The levels of the Caspian Sea during the Upper Pleistocene – the glacial periods – were characterized by the aridization of the catchment area of this lake, which says a lot about the precipitation fell on the basin of the Volga at that time. At the maximum of the last glacial period, c. 18 ka BP, the level dropped to about 60 m below MSL, which is about 32 m below its present level. At 16 ka BP, the level of the lake rose abruptly and, after certain fluctuations, then experienced another severe drop to 58 m below MSL from c. 11 ka to c. 10 ka BP. This episode can be correlated with the cold spell of the Younger Dryas. During this period (the Mangyshlak regression), arid desert conditions prevailed in the Caspian region, creating desert landscapes with eolian landforms. The cold climate is evidenced by pebbles carried by floating ice found in the central part of the lake.

At the beginning of the Holocene, *c.* 9.8 ka BP, the water level rose to 11 m below MSL, indicating the return of a warm climate. This was followed, *c.* 8.5 ka BP, by a cold phase and a fall in the level, reaching as low as 50 m below MSL. From *c.* 8.2 ka to 7.3 ka BP, the catchment region of the Caspian again became humid and, as a result, the level of the lake rose to *c.* 20 m below MSL. From *c.* 7.3 ka to 6.5 ka BP, the water table again dropped, this time to 44.5 m below MSL. The next 1300 years, up to about 5.5 ka BP, were again warm and thus humid, which caused the level of the lake to rise to 25 m above MSL. The ensuing fall in lake level, spelling a return to a cold and dry climate, occurred between 5.2 ka and 4 ka BP. A warm and humid period, extending from 4 ka to *c.* 3.2 ka BP, followed, bringing the level of the lake up to 20 m below MSL. This was interrupted at 3.2 ka BP by a cold period and a low lake level (37.5 m below MSL), which lasted about 200 years. About 1.8 ka BP, there was an abrupt rise in the level of the lake to its former height, followed by a fall from *c.* 1.6 ka to *c.* 1.2 ka BP. This was followed by a short rise, peaking about 1 ka BP. A short low level from *c.* 0.9 ka to 0.8 ka BP followed.

The most recent history of the level of the Caspian Sea consists of small fluctuations (Chepalyga, 1984). While there are differing opinions with regard to the dating of Upper Pleistocene lake level fluctuations, because of technical problems of contamination, there is quite good agreement with regard to the levels for the last 3500 years, as can be seen from the results of the study based on more archaeological and radiometric dating (Mamedov, 1997). In general, his conclusions were that warm global periods indeed coincided with a rise in the level of the lake, while cold periods caused a regression. According to Mamedov, the pattern of transgressions and regressions showed a 450–500 year cyclicity. A more general curve for the oscillations of the Caspian Sea based on geomorphological evidence from deltas and terraces was presented by Rychagov (1997). Although less detailed, it showed a general conformity with the above-mentioned levels.

Comparing the fluctuations in level of the Caspian Sea with those of the Vistula and with climate changes in the Levant during the Holocene (Fig. 2.8), one can see a rather good agreement, particularly in the upper middle part of the Holocene. In the lower part of the Holocene, although the sequence is very similar, there still exist discrepancies of about 500 years in the dates of the changes, which may be a result of some inaccuracy in the dating, as well as of responses of the basin to climatic changes.

3 Climate changes during the Holocene in east Asia (China, Korea and Japan)

3.1 CHINA

Figure 3.1 shows the East Asian area.

3.1.1 Contemporary climate

The contemporary climate of west Asia is dominated in winter by the polar continental air mass (PCAM). During this period there is a northerly flow at the lower troposphere layer, which comes from cold and dry air of middle–high latitude. In the summer, the region is dominated by the tropical–sub-tropical oceanic air mass (TOAM) and the tropical continental air mass: a southerly monsoon dominates the lower troposphere layer, bringing oceanic warm and moist air. There are two types of summer monsoon, the southwestern and the southeastern, influencing different areas. Today, the southeastern monsoon dominates most of China and, in purely theoretical terms, it should have been so for the last 130 ka years (An Zhisheng *et al.,* 1991a).

3.1.2 Climate changes during the Upper Pleistocene and Holocene transition period

During the transition from the Pleistocene to the Holocene, the climate in eastern Asia, especially in China, was different from one region to another, according to the climatic belt to which each region belonged. While the most northwestern region belonged to the westerlies belt, the rest of China was influenced by the monsoon regime. Thus, while the first region was cold and humid during the last glacial period and became warmer and drier as the glaciers melted, most other regions, especially the inner ones, were dry and cold during the Ice Age and warm and moist as deglaciation proceeded (Li Jijun, 1990). This is manifested in the changes in the character of the different deposits. For example, Chaiwopu lake in the Xinjiang region, which is dominated by the westerlies, was 25–28 m higher than its present level depositing gray lacustrine deposits, during the period *c.* 15 ka to *c.* 12 ka BP (Jingtai and Kaqin, 1989). By comparison, the Qinghai and Qaidan lakes in the monsoon zone nearly or entirely dried up during the last glacial maximum (Kezau and Bowler, 1985; Ponyxi *et al.,* 1988). However, when maximum interglacial conditions prevailed, monsoon influences could have also extended to northeastern China, which were usually under the influence of the westerlies (Zhong-wei Yan *et al.,* 1990).

Wang Sumin and Li Jianren (1991) discussed the temporal and spatial distribution of lake sediments in China during the Late Cenozoic and the climatic environments revealed by lacustrine sediments since the last glacial period. They found an impressive variation in the evolutionary features of the lakes in different areas since the last glaciation. The Qinghai and Daihai lakes, lying on the northwestern margin of the east Asia monsoon area, had low lake levels during the last glaciation and the sediments were coarse. The palynological assemblage from Daihai Lake indicates that there was an *Artemisia* steppe, scattered with *Ephedra* shrub thickets and a few conifer or *Abies-Picea–Pinus* trees. The level of the lakes rose from 10 ka to 4.5 ka BP, building a terrace 40–45 m above present level.

3.1.2.a THE LOESS PLATEAU

On the whole, the climate changes during the Pleistocene as well as during the Pleistocene–Holocene transition period are clearly exhibited by the changes from loess to paleo-soil in sediment layers. In the loess plateau of China, influenced by the monsoon, the deposition of eolian loess characterized the dry and cold glacial periods. By comparison, paleo-soils formed during the warm and wet interglacial periods. In the most northerly region, influenced by the westerlies, brown and black soils were formed during the glacial periods, while during the interglacial periods, mainly eolian loess was deposited.

This general pattern varied with the specific local character of the different regions. For example, in the Yulin area, which is on the margin of the desert in the northern part of the monsoonal loess plateau, eolian sand, instead of loess, was deposited during the cold dry glacial period (Guarong, 1988).

An Zhisheng *et al.* (1991a) used the loess–paleo-soil sequence to identify variations in the east Asia monsoon for two time intervals: the last 130 ka years and the last 18 ka years. Here, only

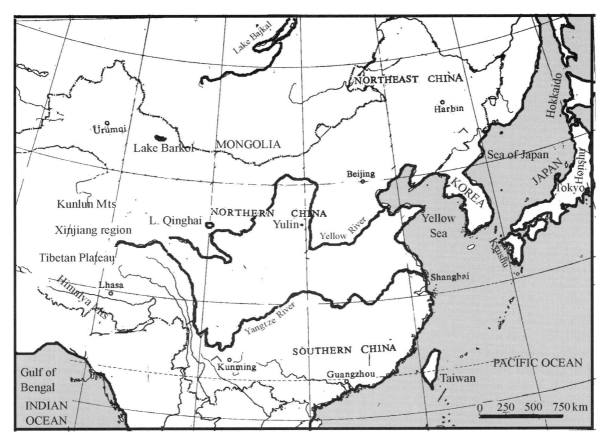

Fig. 3.1. Map of eastern Asia.

the latter, shorter, time span will be discussed. The accelerated accumulation of thick loess layers from 18 ka to 14 ka BP is evidence of the domination of the PCAM or winter monsoon and the decline in the influence of the TOAM, or summer monsoon. The small amount of pollen, dominated by that of herbs resistant to arid conditions, indicates a substantial decrease in rainfall. Around 12 ka BP, a paleo-soil developed in the western loess plateau, there was a significant increase in the pollen of broad leaves trees like spruce and fir of the lakes and the levels rose. All these indicate the strengthening of the summer monsoon. During the transition interval from Pleistocene to Holocene (*c.* 11 ka BP), there was another accelerated accumulation of dust, increase of herb pollen, the disappearance of spruce and fir, and a rapid lowering of levels of the lakes. These data indicate a return to the dominance of the PCAM, which may have lasted for only a few hundred years. Later, the climate changed quickly. From 9 ka to 5 ka BP, a paleo-soil developed on the loess plateau, containing increased pollen of broad-leaved deciduous trees.

An Zhisheng *et al.* (1991b) also summarized the paleo-environmental changes in China during the last 18 ka as evidenced by dust accumulation, vegetation evolution, mountain glacier advance and retreat, and changes in sea level. They used certain magnetic properties of the sequence of loess, related to the precipitation and vegetation coverage ratio, as proxy-data for climate (Kukla and An Zhisheng, 1989; Versoub *et al.,* 1993). From the composite curve, they concluded that the last full glacial age began to decline at about 14.5 ka BP, and there was a cool and humid ripple at about 12 ka BP. The climate then changed rapidly to a cold and dry regime around 11 ka BP and further transformed rapidly to a warm Holocene climate optimum, spanning from 9 ka to 5 ka BP. Afterwards, and continuing to the present, the climate reverted to a cool and dry regime, with occasional ripples of neo-glaciation. The same conclusions were reached by Zhou Weijian and An Zhisheng (1991), who reported on the correlation of ^{14}C chronozones and other indications of the history of climatic changes during the Upper Pleistocene and Holocene on the loess plateau in China. They also considered paleo-soil layers with high "magnetic susceptibility" (MS; i.e., soils with a high iron content) to be indicative of low eolian accumulation and intensive soil-forming processes in which biological activity was involved (pedogenetic processes). Such processes occur most rapidly in a warm and humid climate. By comparison, low MS would indicate rapid dust accumulation and weak pedogenetic processes typical of dry–cold or dry–cool climate. In addition, they made use of the 'magnetic susceptibility age conversion equation' developed by Kukla and An Zhisheng (1989). From 13 ka to 12 ka BP, evidence for a mild climate was found in the form of a carbonate nodule horizon. From 11 ka to 10 ka BP there was a cold and dry climate, while a thick black paleo-soil, indicating a warm humid climate, was formed between 10 ka and 5 ka BP. Since 5 ka BP, the loess plateau has been dominated by the deposition of recent loess, generally reflecting a dry and cold climate. Some intercalations of weakly pedogenetic paleo-soils indicate periods of milder climate.

Hovan *et al.* (1989) established a connection between the Chinese loess sequence and the $\delta^{18}O$ chronostratigraphical and paleo-climatic sequence from a core in the northwest Pacific. Since the formation of the mid-latitude loess occurred during the glacial periods of the Pliocene–Pleistocene Epochs (Pye, 1987), a correlation was observed between the deposition of loess layers in China and greater accumulation of eolian material in the deep sea. From the correlation diagram, presented by Hovan *et al.* (1989), the period of melting of the ice at the end of the last glacial period, characterized by the oceanic water reducing its $\delta^{18}O$ contents, was also distinguished by a reduction in the eolian flux.

Huang *et al.* (2000) have investigated the loess profile in the southern part of the loess plateau. They found that crop cultivation began in this region at *c.* 7 ka BP, yet disturbance of the profile by cultivation has not masked the evidence for eolian dust deposition, which was characteristic of arid periods. During periods of humid climate, fertile brown soil developed. The following is the loess–soil sequence:

before 11 ka BP: semi-desert conditions;
11 ka to 8.5 ka BP: post-glacial amelioration of climate;
8.5 ka to 6 ka BP: humid climate and formation of brown soil;
6 ka to 5 ka BP: arid climate deposition of loess;
5 ka to 3.1 ka BP: humid climate and formation of brown soil;
3.1 ka to 2.2 ka BP: arid climate deposition of loess;
2.2 ka BP to present: humid climate and formation of brown soil.

The rather general character of the section limits the possibility of correlating it with other sections in the region or with conditions in the Levant. In general, it can be said that the dry period spread over time sections of mostly cold global periods, while the humid periods are those that were mostly warm.

3.1.2.b THE TIBETAN PLATEAU AND WESTERN CHINA
Climatic changes in western China have been derived from palynological analysis of sediments, from the mountainous Lake Barkol in eastern Xingjiang (Han Shu-ti and Yuan Yu-Jiang, 1990). These sediments were continuous from the upper Pleistocene to recent times and would be used to examine the effect of

the uplifting of the Tibetan plateau on the climates of western China and eastern China. This uplift, which reached its present elevation of 4000 m during the upper Pleistocene, reduced the influence of the east–western monsoon and resulted in changing the cold–dry/warm–wet to a cold–wet/warm–dry regime. The former, which was consistent and which still characterizes most of present-day China, prevailed in this region only during the Lower and Middle Pleistocene. The uplift also caused the aridization of this region, especially during the interglacial periods. During glacial periods, effective precipitation increased, resulting in lake level rise. During such cycles, low temperatures and an extended icebound season inhibited evaporation, and the greater available moisture caused the forest belt to expand. During the uppermost Pleistocene, most of the period from 21 ka to 12 ka BP was cold and wet, except for three cool and wet periods at 20 ka, 16 ka and 13 ka BP. During these periods, a retreat of the glaciers occurred. At 11 ka BP there was a warm dry spell, but at 10 ka BP the climate became warm and dry.

Li Shuan-Ke (1990) investigated the fluctuations of the levels of closed lakes and the variations of paleo-climates in the north Tibetan plateau, China. He concluded that these lakes went through four fluctuations from 18 ka BP to the present. In the first stage, from 18 ka to 12 ka BP, the level of the lakes was rising, while the mountain glaciers advanced. This was possibly related to the pattern of atmospheric circulation: the belt of westerlies shifted over the region and precipitation increased. In the second stage, from 12 ka to 8 ka BP, the lake levels dropped. During the third stage, from 8 ka to 6 ka BP, a rise in lake levels occurred, while in the fourth stage, from 6 ka BP to the present, there has been a drop of the level of the lakes.

3.1.2.c EASTERN CHINA
Xue Chunting *et al.* (1991), summarizing the history of Holocene coastal sedimentation in China, reported a transgression with a sea depth of 60 m all along the eastern part of the country from 11 ka BP, reaching its climax at 6 ka BP. Since then the coasts have prograded seaward at various rates.

Yang Huai-Jan and Xie Zhiren (1984) reported on sea-level changes in east China over the past 20 ka, relating them to climatic changes. During the Holocene, they correlated minimum sea levels with maximum cold climatic phases observed in China. These were from 8.7 ka to 7.7 ka BP (peak at 8.2 ka), 6.7 ka to 5.7 ka BP (peak at 5.8 ka), 3.1 ka to 2.7 ka BP (peak at 3 ka) and 0.4 ka to 0.1 ka BP (peak at 0.2 ka). During the cold periods, sea levels fell between 2 and 4 m. During the warm periods, a rise was observed. Three major rises could be discerned, from 10 ka to 8.3 ka BP, from 8 ka to 7 ka BP and from 6 ka to 5.5 ka BP. These authors have also investigated the sea-level and climatic changes during the last 2000 years.

3.1.3 Climate changes during the Holocene

3.1.3.a THE LOESS PLATEAU

Kukla and An Zhisheng (1989) reconstructed a MS curve for the Baxia loess plateau. As explained, high MS indicates that soil layers contain higher concentrations of iron oxides, which are produced by soil-forming weathering conditions in a warm and humid climate. In contrast, low MS is characteristic of non-weathered loess, deposited during cold and dry conditions. The MS values showed cold dry conditions between 10 ka and 9 ka BP, as well as c. 7.5 ka BP. From 7.5 ka to c. 5 ka BP, it was warm and wet; from c. 5 ka to c. 3.5 ka BP, it was cold and dry before becoming warmer until c. 3 ka BP, when a colder spell occurred. A further short period of warming was followed by a cold dry spell; a warm period (c. 1.7 ka BP); a relatively cold period; a warm period, which continuing to c. 1 ka BP and was succeeded by a further cold spell.

3.1.3.b THE TIBETAN PLATEAU AND WESTERN CHINA

Ice cores from the Tibetan plateau of China provide a detailed record of the climatic conditions in the period from the Upper Pleistocene to the Holocene (Thompson *et al.*, 1989). The late glacial stage was characterized by lower $\delta^{18}O$ content, increased dust content, decreased soluble aerosol concentrations and reduced ice crystal sizes compared with the Holocene. Between 30 ka and 10 ka BP, concentrations of chloride and sulfate ions gradually increased. One explanation for this increase is that conditions became drier and the extent of salt and windblown loess deposits increased. At c. 11 ka BP, a marked increase occurred, probably reflecting the drying out of the freshwater lake. The ice cores also show pronounced climatic changes during the Holocene. However, the discrete 1000 year time averages of the components of the cores limits the possibility of correlation with other regions. In general, it can be seen that from ca 10 ka to 9 ka BP, the $\delta^{18}O$ was still low, relative to the Holocene record, indicating that the climate was colder than later on. From c. 9 ka to 8.5 ka BP, the $\delta^{18}O$ composition increased, as a result of warmer conditions. This period was also marked by an increase in the salt content of the ice, which means that the lakes and swamps in the adjacent arid areas were drying up. There was a maximum increase c. 6.5 ka BP. From 6.5 ka to 5 ka BP, there was another reduction in the $\delta^{18}O$ composition. The period from 4 ka to 0.7 ka BP was marked by an increase in the $\delta^{18}O$ composition. The Little Ice Age is clearly marked. According to Thompson *et al.* (1989), the values of the $\delta^{18}O$ record of the last 60 years are much higher (which suggests warmer temperatures); the decades with the highest values are the 1940s, 1950s and 1980s. In their opinion, this phenomenon is related to the greenhouse effect.

A high-resolution study of the $\delta^{18}O$ composition of the ice cores from the Tibetan plateau (Yao Tandong *et al.*, 1997) showed that since 1400 AD, there were three cold and three warm periods. The first cold period was from c. 1420 to 1520, followed by a warm period lasting until 1570. The second cold period, which was also the coldest of the three, occurred from 1570 to 1690. This was again followed by a warm period, which lasted until 1770. From then to 1890, there was another cold period. The climate then began to warm up gradually, until it reached its current levels. The $\delta^{18}O$ composition of the ice during the twentieth century was characterized by high values, which these authors believe indicate a trend of global warming. The three cold and warm periods are also apparent from the tree ring record in the Qilan mountains. However, there are some discrepancies with regard to the beginnings and ends of these periods.

Climatic changes in the Tibetan plateau were also investigated by Fu-Bau and Fan (1987). They did this by analyzing the peat and sand gravel layers distributed in the fluvial bog and marsh plain along the mountain foothills and in cross sections near interior lakes, by the study of glacial activity and by the study of landform changes. They concluded that the Holocene started with the Wumandung interval, from 10 ka to 7.5 ka BP, which, according Fu-Bau and Fan, was slightly cold and dry. It was followed by a warm and moist period from 7.5 ka to 5.5 ka BP. This was followed by a warm but dry period from 5.5 ka to 4.7 ka BP. The Qilongduo interval (Yoli period) from 4.7 ka to 3 ka BP was warm and dry, followed by the cold Ice Age I, from 3 ka to 2.5 ka BP. Then came a somewhat warmer period from 2.5 ka to 2 ka BP, followed by the cold mid-neoglacial interval (Ice Age II) lasting until 1.5 ka BP. The Dawelong interval, from 1.5 ka to 0.3 ka BP, was interpreted as having a mild climate. A study of the curve presented by Fu-Bau and Fan (1987) shows that there was a steep rise in temperature c. 1.5 ka BP and a decrease c. 1 ka BP. A dry and cold spell was observed c. 0.33 ka BP.

Lin Zhenyao and Wu Xianding (1987) summarized some features of climatic fluctuations over the Qinghai-Xizang plateau of Tibet. According to their investigations, the climate of this plateau had attained its present general features by the end of the Middle Pleistocene, when the altitude of the area was already about 3000 m, and the average temperatures were still 4–8 °C higher than at present. Today, the average altitude of the plateau is 4000 m. According to peat layers dated by ^{14}C, the annual mean air temperature was 3–5 °C higher than at present during the period from c. 7 ka to 3 ka BP, and especially during the period from 6 ka to 5 ka BP. From c. 3 ka to 1.5 ka BP, the glaciers advanced. However, the climate was not continuously cold. It is obvious that there was a short warm period toward the end of the second century AD, and during the first half of the third century. From the sixth century to the late twelfth AD, there was a markedly warm period and there were many forests around the city of Lhasa. During the seventeenth century AD, the climate became the coldest for the past 1000 years, the mean temperature being 1 °C colder than today.

From the seventeenth to nineteenth centuries AD, there was glacier advance at many sites. Since the middle of the nineteenth century AD, it has become warmer over the plateau. It was warm and wet at the beginning of the twentieth century AD and is now becoming drier. After analyzing historical and meteorological data, these authors arrived at a 2.6–3.6 year periodicity of flood and drought over the plateau, their meteorological and proxy-data pointing to a "quasi biennial pulse" and two principal cycles relative to sunspot maxima, at intervals of about 11 or 22 years. In addition, there are some indications of a c. 30 year cyclicity (Bruckner cycle).

Lin Zhenyao (1990), investigating the climatic changes in the Xizhang plateau in Tibet during the last 200 years, found that major snowstorms occurred 15 times since the nineteenth century AD, of which, those in 1828–29, 1887–88 and 1927–28, were the most severe. From 1950–2000 AD (except for a severe storm in 1967–68), snowstorms became lighter. Droughts and floods have appeared alternately, and now the drought period is gradually getting longer. For example, the rainfall in 1983 was the lowest for the last 100 years. In general, the climate in Tibet is becoming warmer and drier.

Zheng Benxing (1991) studied Quaternary geology in the Kunlun mountains, north to the Qinghai-Xizang plateau of Tibet and south to the Pamir basin. According to ^{14}C dating, he found evidence of neoglaciation from 4 ka to 3.5 ka BP and at 2.7 ka BP. During the Little Ice Age, there were two or three glacial advances. Presently, most glaciers are retreating.

Cui Zhijiu and Song Changqing (1991) investigated the Quaternary periglacial environment in China. During the period of transition to the Holocene, 13 ka to 10 ka BP, it was drier and colder than today. The period from 10 ka to 9 ka BP then moved to a cool-to-temperate semi-arid climate; from 9 ka to 8 ka BP, it was cold and dry; from 8 ka to 6 ka BP, it was warm and semi-humid; from 6 ka to 5 ka BP, conditions were dry and cold; from 5 ka to 3 ka BP, they were warm and semi-dry; from 3 ka to 1.5 ka BP, they were relatively cool and wet; and, finally, from 1.5 ka BP to the present, it was semi-arid and cold.

Zhou Yuowu et al. (1991) reported on the changes in the permafrost in China during the Quaternary. During the Holocene, from 10 ka to 3 ka BP, the air temperature in eastern China was 3–5 °C higher than the present. At 7.3 ka BP, the air temperature was 1 °C warmer in the Tienshan plateau and 2–4 °C warmer on the Qinghai-Xizang plateau. At about 3 ka BP, it again became cooler and, after a series of fluctuations, the present pattern of permafrost distribution was formed.

During the Holocene, the levels of the lakes in northwestern China followed the pattern that was described earlier for the Upper Pleistocene. For example, several lakes that lie in the eastern Kunlun and Xijang deserts, and which are under the influence of the westerlies, maintained a high water level up to 12 ka BP

Then, the water level fell until 8 ka BP, after which it rose again, attaining its highest level from 6.5 ka to 6 ka BP and then descending sharply (Jijun, 1990).

Wang Sumin and Li Jianren (1991) reported on a general retreat of the levels of the Qinghai and Daihai lakes from 4 ka to 3 ka BP; these lakes lie on the northwestern margin of the east Asia monsoon area. Lakes on the Yunnan plateau, situated in the southwest monsoon area, have had opposite trends of evolution. During the last glaciation, they formed large water bodies, but they retreated during the Holocene. In Xinjiang Province, which is influenced by the westerlies, the lakes were high during the last glacial period (as well as the Holocene optimum) but regressed during the past 2000–3000 years. In the middle and lower reaches of the Changjiang river, lakes disappeared during the last glaciation and were replaced by a fluvial plain, filled with loess-like deposits. Lake Taihu came into existence c. 2.5 ka BP, while lake Poyang was formed during the Han Dynasty, c. 2 ka BP.

3.1.3.c CENTRAL AND EASTERN CHINA

Zao Xitao et al. (1990) reported on Holocene stratigraphy and its reflection of climatic and environmental changes from a section in a 8 m thick peat layer in Qingfeng, Jinahu County, of Jingsu Province. They analyzed and carbon dated the sedimentary sequence, the fauna, and the flora, arriving at the following stratigraphy from bottom to top.

Phase 1: peat swamp of the Pre-Boreal period. The vegetation was then coniferous and of steppe type. The climate was arid and cold.

Phase 2: littoral swamp of the Boreal period (c. 9.5 ka to 7 ka BP). The vegetation was salt meadow and needle and broad-leaf mixed forest; the climate was dry and warm–temperate, and seawater affected the area for a time.

Phase 3: lagoon of the Early Atlantic period (c. 7 ka to 6 ka BP). The vegetation was saltwater marsh and pine-bearing plus deciduous broad-leaf forest. The climate was semi-humid and warm; seawater invaded this area, forming a lagoonal environment in the middle to lower part of the inter-tidal zone. During the Atlantic Period the sea level was about 2 m higher than that at present.

Phase 4: bay of the Late Atlantic period (c. 6 ka to 5 ka BP). The vegetation was saltwater marsh, with mixed evergreen and deciduous broad-leafed forest; the climate was humid and warm, and the Holocene transgression reached its maximum. The area studied became an open bay environment in the middle to lower part of the inter-tidal zone to the upper part of the sub-tidal zone. The seawater was warmer.

Phase 5: littoral lowland of the early Sub-Boreal period (c. 4 ka to 2 ka BP). There was a stage of low water levels.

At *c.* 2.4 ka BP, the vegetation was salt meadow and deciduous broad-leaf forest. The climate was semi-arid and warm-temperate, and seawater retreated from this area for a time.

Phase 6: desalted lagoon of the late Sub-Boreal period. The vegetation was mixed needle and broad-leafed forest; the climate was humid and cool–temperate, and seawater again affected this area.

Shi Xingbang (1991) reported on the natural environment of the Neolithic period in China. He divided this age in China into three main cultures.

1. Paddy growing in east China and south of the Yangtze River. The deposits from 7 ka to 6 ka BP (Hemudu culture) bear evidence of an agricultural tribe and contain an assemblage of pollen indicating a moist and warm climate, temperatures being higher than at the present. From 6 ka to 5 ka BP (Qingdun and Songze culture), the climate was an intermediate one between the sub-tropical and temperate zones. It was warm and moist, and lakes and swamps were widely distributed. The inhabitants farmed on the terraces and also hunted and fished. From 5 ka to 4 ka BP, the climate became dry and cool; the sea level went down, and people lived mainly on agriculture. In the Yangtze delta area, the change of climate and sea level controlled the development of settlements. From 8.5 ka to 7.5 ka BP, the climate became cold; the sea level was 10 m lower than at the present, and there were no human activities. From 6.5 to 5 ka BP, a delta was formed to which people migrated. From 5 ka to 4 ka BP, the climate became dry and cool, and the area of water was reduced.
2. The millet type agriculture on the loess plateau. From 8 ka to 7 ka BP (the pre-Yangshao culture), the climate was warmer than at present, temperatures being 2–3 °C higher. From 7 ka to 4 ka BP (The Yangshao culture), the climate was cooler and drier in the earliest part, warm and humid in the middle period and dry and cold towards the end.
3. The microlithic culture of northeast China, inner Mongolia and Qinghai-Xizang plateau. From 10 ka to 8 ka BP, an arid forest prairie dominated; 6 ka to 4 ka BP was the warmest period, the temperatures being 2–4 °C higher than today. From 4 ka to 3 ka BP, the climate was arid; lakes shrank, the prairie enlarged and salinization intensified.

Zhang Peiyuan and Wu Xiangding (1990) investigated climate fluctuations in China based on changes in crop patterns, as reflected in archaeological and written records. They found that millet was the main crop mentioned from 8 ka to 3 ka BP but later on, during the Shang Dynasty, broomcorn millet was often depicted on oracle bones. Little is mentioned about other crops, which implies that, by the era of the Yin people, broomcorn

millet was the main cereal cultivated throughout the Yellow River valley. Since broomcorn millet needs a shorter growing season than millet, there may have been a sharp drop in temperature and humidity from 3.4 ka to 3 ka BP After this cold period, the climate warmed up again and millet was mentioned more frequently. During the Warring States period (402–401 BC), people lived on millet and beans. The cultivation of broomcorn millet shifted or migrated northward.

Another indicator of a mild climate in historical times, according to Zhang Peiyuan and Wu Xiangding (1990), is the plum tree (*mei* in Chinese). This tree was quite widespread in the Yellow River valley from the tenth to sixth centuries BC but practically disappeared from the Yellow River after the Song Dynasty (960–1279 AD). The cold climate during the twelfth century also influenced the cultivation of lychee, a typical Chinese tropical fruit, widely grown at least since the Tang Dynasty (618–907 AD) in southeast China. There, lychee harvests were destroyed twice by cold weather, in 1110 and 1178 AD. By comparison, according to official histories, the thirteenth century was a very warm period, during which southern winds prevailed, attesting to mild monsoon weather. This caused a shift to the north of the boundary of the pasture and mixed agriculture plant system, relative to the present, and the establishment of many agricultural settlements that do not exist today. This northern boundary coincided with the 400 mm rain line. Thus, at that period, this line shifted to the north by 50–80 km. From 1302 AD onwards, much damage from cold weather was reported, marking the end of the warm period and the first stages of the Little Ice Age.

Zhang Peiyuan and Wu Xiangding (1990) examined the influence of the Little Ice Age by determining the number of severe winters per decade, which could be deduced from historical records of large frozen rivers, lakes and wells. These show that from 1500 to 1978 AD, there were 136 severe winters. These were especially bitter in 1500–50, 1601–1720 and 1831–1900 AD. The periods of warmer winters were 1551–1600, 1721–1830 and 1901–50 AD.

According to Zhang Peiyuan and Wu Xiangding (1990), the variety of data from ancient documents, especially by government officers who prepared reports for the emperor, provides a rather detailed picture of the climate of the eighteenth century. These were correlated with measured data for the period 1960–90 and, thus, temperatures for the eighteenth century were obtained. The percentage of snow and precipitation days in the 1720s and 1770s was 10–15% above that of today, and the derived temperature was 1–1.5 °C lower than that of today. Therefore, although the eighteenth century is regarded as a warm period within the Little Ice Age, these results show that it was colder than at the present. Records of the dates at which certain flowers bloomed suggest that these were, on average, 6 days later than today and, accordingly, it can be estimated that the climatic belt has moved southward by

1.5° in latitude. Comparing observations regarding the directions of wind made from 1723 to 1769 AD with those of today, it can be inferred that the wind directions have also changed. In addition, the analysis of the records showed that the length of the Beijing rainy season has extended during the last 274 years, the span from 1947 to 1978 being the longest. The shortest season occurred in 1746 AD, and there were relatively dry periods in 1724–63, 1814–63 and 1924–43 AD.

Chinese provincial historians recorded many bad years caused by floods and droughts during the last 500 years. Zhang Peiyuan and Wu Xiangding (1990) and Zhang Peiyuan and Ge Quansheng (1990), analyzed these records and found that the main abrupt events of drought/flood disasters occurred in five time periods: 1500–50, 1600–50, 1720–40 and 1900–40. Since most abrupt events in this study were related to warming trends, it may be deduced that the greenhouse effect may also increase the instability of sensitive areas to drought/flood disasters.

The same authors also used historical data to analyze the impact of climate changes on farming systems. Rice and wheat farming was practiced in the Zhejing and Jiangsu provinces (lower valley of Chang Jiang) after the twelfth century. When the climate had deteriorated by the seventeenth century, both wheat and rice had difficulty ripening, and the dates of sowing wheat were shifted to earlier in the year in order to protect it from winter damage. Cold climate again affected the yields of wheat and rice by the end of the eighteenth and the beginning of the nineteenth centuries. Wheat and cotton farming, introduced to the middle and lower valleys of Chang Jiang during the Ming Dynasty (fourteenth to seventeenth centuries AD), suffered from the cold. In 1628 AD, the farmers were advised to sow their wheat before the cotton was harvested in order to avoid winter damage. The double rice crop was also negatively affected when the climate became colder.

Citrus fruit are highly susceptible to cold climates and can be used to monitor climate changes. As source areas of citrus fruit sent to the emperor as tribute were recorded, it can be seen that the places which supplied citrus fruits in the seventh to tenth centuries AD were less affected by frost. This would suggest that the climate was warmer during these years.

The distribution of hemp was also used to detect climate change. From this data, the same authors concluded that the thirteenth century AD was the warmest period of the last 1000 years.

Zhang Peiyuan and Wu Xiangding (1990) also investigated the paleo-climates of the Tibetan plateau. They suggested five main divisions.

1. 7 ka to 3 ka BP: climatic optimum. It was warm and wet; annual mean air temperatures were 2–3 °C higher than those of today;

2. 2.9 ka to 1.6 ka BP: neo-glacial period. It was cold, with frequent large glacial advances;

3. 1.5 ka to 0.9 ka BP: warm period. Climate was warmer and wetter than at present;

4. 0.89 ka to 0.15 BP: Little Ice Age. It was colder, especially in the mid seventeenth century, about 1 °C lower than at present;

5. 0.14 to 0 BP: last warm period. It was warm and wet at the beginning of the twentieth century, becoming progressively drier.

Based on historical official files, these authors found that floods and droughts appeared alternately during the last 100 years, and there were three rainy periods (1883–1906, 1916–34 and 1947–62) and three dry periods (1907–15, 1935–46 and 1963–80). They conclude that the Tibetan plateau is an area more sensitive to climatic change than other areas at the same latitude.

In addition to the observations described above, Zhang Peiyuan and Wu Xiangding's (1990) main conclusions were that, although individual catastrophes may occur during periods of warming, the history of China shows that cold climates have had a more serious effect on agriculture than have warm periods.

Wang Jian et al. (1990) developed an original method of "factor interpretation" based, in the first stage, on factor and correspondence analyses; these enable the principal factors controlling pollen changes in a fossil pollen sequence to be identified and the relationships between these factors and the pollen taxa to be calculated. The second stage determines which one of these relationships is the climatic factor. This allows a weighting sequence to be constructed for the climatic factors, in terms of the relationship between the climatic factor and the samples. In the third stage, these weightings are transformed into parameters (temperature, precipitation, etc.) in relation to ecological features. This method has been applied to peat deposits in three Holocene lakes, two of which are in northeastern China (Sanjiang plain and Mount Changbaishan) and one is a small lake in southwestern China (Mount Loujishan). The results suggest that the period from 12 ka to 10–9 ka BP was one of rapid warming. Yet, the summer half year temperature (SHYT) was about 8 °C lower at 12 ka BP than at the present. However, from c. 10 ka to 9 ka BP, it reached present day temperatures. The average SHYT from 9 ka to 3 ka BP was 2–3 °C higher than today, the thermal maximum, with a SHYT 3–4 °C higher than today occurring between 7.5 ka and 6 ka BP in north-eastern China. There was an obvious deterioration of climate from 3 ka BP. Seven periods of cooling were observed to have occurred during the Holocene: from c. 9.1 ka to 8.8 ka BP; from c. 8.3 ka to 8 ka BP; from 5 ka to 4.7 ka BP; from c. 4.4 ka to 4.1 ka BP; from c. 3 ka to 2.7 ka BP; from c. 1.8 ka to 1.4 ka BP; and from c. 0.5 ka to 0.3 ka BP.

The palynological study of a core from lake Barkol in Xinjiang (mentioned above; Han Shu-ti and Yuan Yu-Jiang, 1990) gives a rather detailed picture of the climatic changes that took place in most of western China; that is the regions shielded from the monsoons and thus under the influence of the westerlies belt. Looking at pollen data using ^{14}C dating for the various levels in the core from the bottom (earliest) up, it can be seen that, the climate at c. 12.5 ka BP (layer C6) was cold–wet; lake levels were high and the glaciers were extended. By c. 10.87 ka BP (layer W6), conditions had reversed; it was warm and dry with low lake levels and retreat of glaciers. Above this, there was a short cold–wet spell indentified within the same layer (C5). From 9.37 ka to 8.97 ka BP (layer W5–2) and at 8.19 ka BP (layer W5–1), the climate was warm–dry. At 7 ka BP (C4), it was cool and wet; from 6.618 ka to 4.18 ka BP, the layer (W4) is interpreted by these authors as warm–wet, but they mark a cold–wet climate and high lake level at 5 ka BP (C3). From 4 ka to 3.64 ka BP (layer W3), the climate was again warm–dry; the lake dried up completely and the glaciers retreated to above 3530 m. From 3.27 ka to 2.64 ka BP (layer C2), the climate again became cold–wet; the lake level was high and the glaciers descended to 2830 m. From 2.31 to 1.2 ka BP (layer W2), the climate reverted to warm–dry, with low lake levels. In layers consisting of the upper 0.36 m of the section, corresponding to the seventeenth to nineteenth centuries, there were three alternating periods of cold–wet and warm–wet.

Han Shu-ti and Yuan Yu-Jiang (1990) also observed that all historically noted disastrous weather conditions occurred during the transitions from cold to warm or warm to cold periods. For example, almost all heavy snow events in the Barkol region during the last 300 years (1679, 1867, 1882, 1934, 1951 and 1973) happened during such changes. Based on tree rings and precipitation records, these authors observed a marked warming and drying tendency in the Xingjiang region. Mean precipitation declined by 12.5% and the frequency of the occurrence of drought years increased by nearly 100% from the eighteenth to the nineteenth century, while wet years declined to one seventh of those of the eighteenth century. The data from 13 weather stations in Xingjiang region showed that the temperature was rising. Based on these observations, Han Shu-ti and Yuan Yu-Jiang, (1990) forecast that climate warming would have a very negative effect on this region.

Sun Xingjun and Chen Yinshuo (1991) summarized the palynological research carried out during the previous 30 years in China. Their main conclusions were as follows.

1. Around 10 ka BP, the changes in vegetation to those more suited to warm conditions were sudden and remarkable. This change took only a few hundred years.
2. There was a warm Holocene period between 10 ka and 4 ka BP. This period can be divided into sub-periods. From 10 ka to 8 ka BP, temperatures were rising, and from 8 ka to 4 ka BP was a period of high temperatures.
3. The temperature decline in the Late Holocene, usually considered to start at 2.5 ka BP, may have started earlier, in most cases c. 4 ka BP. In contrast to the change at the beginning

of the Holocene, this alteration was much slower and more gradual, taking some 1000 to 2000 years to stabilize.

Wen Qizhong and Qiao Yulo (1990) investigated the climatic sequence in the last 13 ka in the Xingjiang region. They based their study on "dated sediments of Lake Aibi, the loess layers in the northern foothills of Tienshan Mountains and dune sections of Mosuowan and Damagou, as well as the river lake sediment section of Yuetegan in Hetian country". Their sequence of climatic changes in this region is as follows:

1. 13 ka to 10 ka BP: cold–moist to cool–dry;
2. 10 ka to 8 ka BP: warm–dry;
3. 8 ka to 7 ka BP: warmer–moist;
4. 7 ka to 6 ka BP: temperate–dry;
5. 6 ka to 4 ka BP: warmer–moist;
6. 4 ka to 3 ka BP: warm–dry;
7. 3 ka to 2 ka BP: cold–moist;
8. 2 ka BP to present: warm–dry (with a short episode c. 1.5 ka BP that was cold–moist).

Yang Huaijen (1991) investigated the relations between the paleo-monsoon and the mid-Holocene climatic and sea-level fluctuations in China. Yang Huaijen and Chen Xiqing (1987) found evidence showing that there was an encroachment by the sea c. 8 ka to 6 ka BP, which inundated the north China plain and the Yangtze delta plain. The air temperatures attained a climax c. 7 ka BP, but the climax of the sea transgression occurred about 1000 years later. (At c. 6 ka, 5 ka and 4 ka BP, great floods also occurred over the flood plain of the Yangtze river because of increased discharge.) From 5.6 ka to 5.3 ka BP, low temperatures and low sea levels persisted for some time. There are indications that, from c. 4.6 ka to 3.8 ka BP, a period of high sea levels occurred synchronously with heavy rainfall and high temperatures. According to Chinese historical documents, prolonged heavy rainfall and great floods occurred in North China in the dynasties of Yao, Shun and Xia (4.4 ka to 4.1 ka BP): the most widely spread flood legend among the Chinese peoples. According to ancient Chinese documents, many rivers overflowed their banks and spread unchecked. Some tributaries of the Yellow River reversed their courses as a result of elevation of the base level and silting up of their lower channels. Because of the floods, people moved to higher ground and began to learn to dig wells.

Li Pingri and Fang Guoxiang (1991) investigated the Quaternary deposits and environmental evolution in the Guangzhou plain and Zhuziang delta in southeastern China. They analyzed sporo-pollen assemblages from deposits in this region and found that the climate in the Early Holocene was generally warm and wet, with some colder periods, turning to warm–dry in the later period of the Holocene. They found the Upper Holocene to have been warm and dry from c. 4.5 ka to 5 ka BP, hot and wet from c. 4.5 ka to 3.5 ka BP, warm and somewhat drier from *c.* 3.5 to 3 ka BP and hot and wet *c.* 3 ka to 2.5 ka BP.

A preliminary study of the climatic sequence from 7.5 ka to 5 ka BP of the middle and lower beaches of the Yangtze River was based on calculations of temperature from palynological data from 50 sites (Tang Lingyu *et al.,* 1990). The calculations were based on the weighted distribution statistical formula. The results suggested that the average annual temperature from 7.5 ka to 5 ka BP was 0.1– 2.71 °C higher than at present; the hottest period was between 6.5 ka and 6 ka BP, when the mean temperature was 1.7 °C higher than it is today; the coldest period was from 5.5 ka to 5 ka BP, when the mean temperature was 0.4 °C higher than today.

Fang Jin-Qix (1990), citing data from Chinese historical documents, deduced that several sharp temperature fluctuations occurred in historical times. The first cold phase was around 3 ka BP, with temperatures 1–2 °C cooler than those of today, and must have lasted more than two centuries. The second cold period occurred from 100 to 600 AD, the third, from 1050 to 1350 AD and the fourth from 1600 to 1850 AD, with coldest temperatures occurring around 1700. The warm spells took place from *c.* 700 BC to the end of the millennium, 650 to 1000 AD (during this period mean temperature was 2–3 °C above that of the present) and also *c.* 1400 AD. In colder climatic periods, usually rare winter thunderstorms also appeared much more frequently. These were associated with rare frontal movements of very cold air masses from central Asia. Such events occurred during 100–150, 300–500, 1100–1300 and 1600–1900 AD. Also, there were frequent dust storms during the cold periods. The longest break without electrical winter storms was from 600 to 800 AD.

Based on the analysis of a drought-flood index, five drier climatic intervals lasting longer than 100 years were identified in northern China since 2 ka BP (Fang Jin-Qix, 1990, citing Zheng, 1985). These were 0–100, 300–630, 1050–1270, 1430–1550 and 1580–1720 AD. The driest climatic periods in northern China, especially in the region from latitude 30° to 45° N, coincided with colder climates. This can be explained by strengthening of the Siberian anticyclone and southward shift of the polar front. It was wetter in the intervening periods, the longest such period since 0 AD being from 810 to 1050 AD. The records of the floods of the Yellow River show three intervals of low frequency, which more or less coincided with colder and drier periods: 150–650, 1120–1280 and 1680–1880 AD. According to Fang Jin-Qix (1990), the longest break with no floods occurred from 153 to 637 AD, leading to the desiccation of 29 lakes, of which 27 totally disappeared.

Fang Jin-Qix (1990) found a close correlation between climate changes and the migration of Chinese peoples in historical times (Fig. 3.2). His study was based on a critical survey of historical data on these migrations, as well as a literature survey of the information about Chinese lake evolution, the history of agro-cities in the Chinese deserts and flood reports of the Yellow River. The

main migratory events coincided with social and political unrest, as well as with invasions from the north by the Mongols. All these events, according to this author, were connected with severe climatic changes. According to Chinese researches (Hanyong and Shanyu, 1984; Wenlin, 1988), the percentage of Chinese population settled in southern China as a result of these events grew from about 17% of the total Chinese population 2000 years ago to about 65% at the beginning of the nineteenth century.

According to Fang Jin-Qix (1990), these Chinese migrations were in phase with the cold climate phases, which occurred *c.* 1000 BC, 0–50, 300–600, 1100–1300 and 1600–1750 AD (Fig. 3.2). The first recorded migration at 1000 BC was that of the Zhou (Chou), who were western nomads who settled in present day northwestern China, moved eastward, overthrew the rule of the Shang Dynasty and established the Western Zhou Dynasty. During this period, the glaciers advanced and the snow line descended by 100–300 m in the mountains of Tibet. Some sand dunes were also reactive and some lakes in the west and north of China became more saline.

During cold and dry periods, there was also a southward shift of the agro-husbandry boundary. For example, in 300–500 AD, the boundary was 200–400 km further south of that in either the former or the subsequent warmer and more humid periods.

According to Fang Jin-Qix (1990), an analysis of the records of productive outputs and rainfalls for the previous 30 years in the Inner Mongolia Autonomous Region suggests that there is a linear relationship between summer precipitation and herbage outputs. Low summer precipitation causes the growing period of the vegetation to shorten, thus reducing the carrying capacity of the plains and leading to famine among livestock and the dependent population. The general conclusion of this author is that the greenhouse effect should be positive for China's agriculture as it would strengthen the monsoons, causing agricultural regions of China to become more warm and humid and thus productive.

Zhang Lansheng *et al.* (1990) reported on severe droughts that have occurred since the beginning of the nineteenth century. The worst was from 1920 to 1929, reaching a maximum in 1928 and 1929. During this period, some main tributaries of the major rivers of northern China, including the Yellow River, dried up, as did inner land lakes in Inner Mongolia. Wells and springs failed and sand desertification extended. At the same time, in northern China, floods, frost, hail and war caused serious famine. Almost no crops were harvested. The famine victims had to eat the bark of trees, roots of grasses, soil and even human corpses. In 1929, more than 6 million people died just in the northern provinces. A lot of people migrated from the region; many women were sold to other areas to slave-traders. The most serious disasters were in the provinces of Gansu and Inner Mongolia, followed by Shaanxi and Henan, Shanxi, Hebei and Shandong provinces.

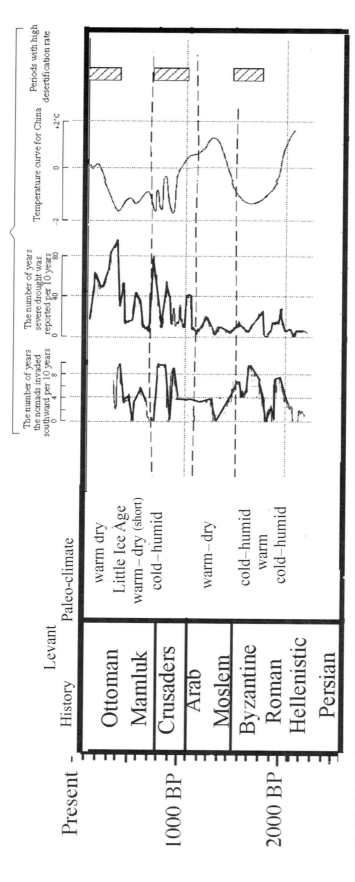

Fig. 3.2. Holocene climates in China. Changes over the last 2000 years (modified from data by Fang Jin-Qix, 1990).

As the previous severe drought had occurred between 1876 and 1878, causing the death of about 13 million people in northern China, while the most recent was in the 1970s, Zhang Lansheng *et al.* (1990) came to the conclusion that there is a 50 year cyclical occurrence of droughts. Each 150 years, the authors maintain, there are chances of a severe drought, like that of the 1920s.

Guo Qiyun (1990) analyzed the possibility of a common cause (tele-connection) for the floods and droughts in northern China and the Indian summer. This analysis utilized a dataset covering 116 years from 1871 to 1986. The correlation coefficient (0.39) between floods and droughts was significant, with a 99% confidence level, although unstable (0.15 for 1921–50 and 0.58 for 1951–80). The variations, both in floods/droughts over northern China and in Indian summer monsoon rainfall, are closely related to the general atmospheric circulation, especially to the Indian Low and the Western Pacific High. When the northern border of the Indian Low moved further north, the correlation between the floods and droughts over northern China became insignificant. At that time, the effect of the Western Pacific High on the floods and droughts in northern China was predominant. A correlation was also found between less summer rainfall over northern China and high sea-surface temperatures of the eastern equatorial Pacific, in summer, autumn and winter.

Interpreting records of autumn rains in Hunan Province during the last 1300 years, Xiong Dexin (1990) found a remarkable periodicity of 90 years, and that changes in autumn rain, with cold to warm fluctuations, were closely related. In the warm periods, autumn rain was lower, while in the cold period it increased. Since the late fifteenth century, autumn rains have shown a clearly increasing trend. Based on predicted higher temperatures, the author forecast a future reduction in autumn rains, spelling an increase in autumn droughts for the Hunan Province. Xu Xiangding and Yin Xungdang (1990) carried out a study of the abrupt dryness–wetness changes in the central Yellow River valley during historical times. They analyzed data derived from tree rings for the last 280 years and from chronologies for the last 760 years. Abrupt changes were found for the end of the thirteenth, the middle and end of the fifteenth, the middle of the seventeenth, the early nineteenth and the early twentieth centuries.

Wang Shaowu *et al.* (1987) investigated the drought-to-flood variations for the last 2000 years in China compared with global climatic change. Based on historical documentation, they extended the drought-to-flood series back to 951 BC, especially for the Changjiang and Huanghe rivers. Generally, they found four dry periods in the Huanghe River, which were during the fourth to sixth, eleventh to thirteenth, fifteenth and nineteenth to twentieth centuries AD. Four dry periods were observed in the Changjiang River, in the fourth, ninth, eleventh to thirteenth, and fifteenth to seventeenth centuries AD. Analyzing the special conditions of the Little Optimum (eleventh to twelfth century AD) and the Little Ice Age (1550–1750 AD), they found that the first period was dry while the second was wet. This is in agreement with information from Japan (Yoshino, 1978), which indicated that the Little Ice Age was cold and wet. This climate is explained by the cyclonic trough in east Asia, which may have been located near Lake Baikal during the Little Ice Age while the blocking Okhotsk anticyclonic high occupied the east coast of Asia.

The climatic characteristics of the Little Ice Age in China were investigated in greater detail by Wang Shaowu (1990). This study was based on historical documents, from which information on early first frost, late last frost, summer snow and bitter cold in summer were calibrated with temperature observations, using decade mean seasonal temperatures anomalies estimated from the frequency of the events. Finally, a series of decade mean annual temperature anomalies was reconstructed. During the last 600 years, three cold periods were identified, each of them consisting of two colder phases. The mean temperature below the average for 1880–1979 is given in parentheses.

Phase I: 1450s to 1470s ($-0.30\,°C$) and 1490s to 1510s ($-0.34\,°C$)
Phase II: 1560s to 1600s ($-0.47\,°C$) and 1620 to 1690s ($-0.60\,°C$)
Phase III: 1790s to 1810s ($-0.43\,°C$) and 1830s to 1890s ($-0.45\,°C$).

The annual temperature average for the coldest phase (1620s to 1690s was 1 °C colder than the twentieth century average. Warming occurred from the 1920s to the 1940s. Although three cold periods were also identified for Europe and North America, there were large discrepancies between China and these countries in the timing of these cold periods; those in China seem to have occurred earlier.

Sultan Hamid and Gaofa Gong (1990) investigated the influence of climate change on peasant rebellions in the seventeenth century, at the end of the Ming Dynasty. The years 1627–43 were characterized by droughts of unprecedented duration over most of northern China. A year-by-year comparison of the geographical distribution of rebel activity with the moisture conditions shows that 75% of the rebellions occurred in drought areas, 15% in areas affected by flood and only 10% in areas with normal weather.

Zhong-wei Yan *et al.* (1990) investigated the northern hemispheric change of climate conditions in the 1960s. Their conclusions vis-à-vis precipitation and temperatures, in addition to atmospheric pressures, were that rainfall in the zone stretching from north Africa to China–Japan decreased abruptly, while that in the zones both to the south and north increased. In addition, temperatures decreased abruptly in most of the northern hemisphere and increased in some low latitudinal regions.

In contrast to Li Jijun's (1990) forecast of a positive influence of global warming on precipitation in China, Fu Congbin *et al.* (1990) forecast a dry trend. They based this forecast on identifying a trend that has developed during the past 100 years. They applied a linear fitting to records of observed drought and wet indices, summer monsoon rainfall, length of plum rains (Meiyu) and the discharge of the Yangtze River since 1887. Superimposed on this long-term trend, there was an oscillation with a period of about 36 years (the so called Bruckner cycle), which is supposed to be related to solar activities and air–sea interaction. The most interesting feature is that the climate abruptly entered a dry regime in the 1920s, as detected by the Mann Kendall rank statistics test. According to their analysis of the meridional profile of the global zonal mean precipitation anomalies during the peak period of global warming (1930–40), the changes of mean meridional circulation would support a relationship between warm climate and dry years: "There would be an enhanced Hadley circulation in the Northern Hemisphere during the warming episode, which has an enhanced descending branch over the sub-tropics and extra tropics where the most part of China is located, while over the tropical region, an enhanced rising branch of the Hadley circulation produced more precipitation over there, which was proved by the enhanced Indian monsoon activities" (Fu Congbin *et al.*, 1990). Their conclusion was that if global warming was to continue, the dry trend in China would continue and be further enhanced in the future.

Pan Tiefu (1990) predicted a similar trend to warmer and drier conditions in the Jilin Province of China. This was based on an analysis of climatic data for Changchun since 1909. The records show that the temperatures in Jilin have gradually risen since 1909. The average yearly atmospheric temperature rose by $0.0097\,°C$; that of May to September, which is the plant growing season, by $0.0026\,°C$, and that of winter, December to February, by $0.0183\,°C$. Pan Tiefu (1990) forecast that, by the middle of the twenty-first century, the average atmospheric temperature of the year would be $0.6\,°C$ higher than today. As the climate of the Jilin province is cold, a trend to warmer conditions may allow an earlier seeding time, which may benefit the agriculture of the province. However, records also show that the annual average rainfall has been reducing since 1909 at a rate of $1.287\,mm$ per year (that from May to September by $0.187\,mm$, and that of December to February by $0.155\,mm$). It is, therefore, predicted that the semi-humid region in the middle part of Jilin Province will gradually change to semi-arid.

Liu Weilun and Wang Yunzhang (1990) found a significant negative correlation between the runoff of the Shanmenxia tributary of the Yellow River and the size of the Antarctic ice cap. Using annual runoff data for the years 1700 to 1988, they found an interchange between floods and droughts. Although there was a period (1922–32) with severe drought, a trend towards long-term drought could not be seen. They found "a significant negative correlation between the runoff and the ice amount, when the runoff is lagging one year behind the ice amount. The correlation coefficient is -0.41, which is significant at the 0.1% level. Therefore, the valley

is likely to be drier as the ice amount increases largely and vice versa. The driest period (1922–1932) of the valley coincided with the peak period of the ice amount". The connection occurs because of the influence of the ice cover on the paths of the Antarctic storms during winter. "When the ice cover is heavy, the cyclonic activities at high latitudes decrease and move equator-wards. Following it, the sea-surface temperature starts to fall in the eastern tropical Pacific Ocean, the Hadley Circulation weakens and the Walker Circulation intensifies, which cause the change of the highs."

Zhang De'er (1991) also reported on climatic changes in the last 1000 years in China based on his interpretation of historical records. He concluded that the mean annual temperatures during the twelfth, thirteenth, seventeenth and nineteenth centuries were the coldest for the last 5000 years. In 1111 AD, for example, the $2250\,km^2$ Thaihu Lake was completely frozen; the ice was sufficiently solid as to be able to bear traffic. In 1110 and 1178 AD, the winter frost killed all the lychee fruit trees. From 1131 to 1260 AD, the 10-year average date of the latest snowfall day in Hangzhou was 9 April, which was 1 month later than that of the twentieth century. In other words, the temperature in April in Hangzhou was $1–2\,°C$ lower than it is today. The snow lines in the Tienshan mountains in the twelfth and thirteenth centuries were possibly $200–300\,m$ lower than today. The fourteenth century was also colder than the present one. In the winters of 1329 and 1353 AD, Taihu Lake was so frozen that it could be walked over, and all the tangerine trees were destroyed by the cold. None of these phenomena have been seen during the present century.

The temperature variations in the recent 500 years display three dominant cold–warm cycles with a periodicity of about 180 years. The three cold intervals were 1470–1520, 1620–1720 and 1820–90. The warm ones were 1530–1620, 1730–1810 and from 1900 to the present. (Zhang De'er did not mention the frosts of the 1920s.) Investigating the mean annual temperatures along the Yangtze River, Zhang De'er came to the conclusion that there had been no warming trend during the nineteenth century compared with figures for the last 500 years (variation range $0.6\,°C$ between winter maximum ($1.8\,°C$) and winter minimum ($1.2\,°C$), while during the twentieth century the range was $0.6–0.7\,°C$), and there has been no general warming in the last 20 years of the century compared with the century as a whole (highest winter temperature for Shanghai and Hankou for the 1940s was 5.0 and $5.1\,°C$, respectively whereas during the 1970s to 1980s, it was 4.9 and $4.7\,°C$, respectively).

Zhang De'er (1991) also suggested using the historical records of spring and autumn frosts. Although these could not be used to infer the spring and autumn temperatures, they could be used to mark the character of cold air activity in spring and autumn. He noted that many annual first frost dates recorded in historical literature of the last 600 years were earlier than, and last ones later than, the present. His statistical analysis showed that the years of killing frost come in clusters. (For example in the Yangtze River basin, they were 1420–60, 1500–40, 1620–90, 1720–80 and 1830–1910. From data reconstructed from the *Qingyulu* (a daily weather record in the Qing imperial archives) and data from modern observations, a mean July temperature sequence for Beijing was obtained for the last 260 years. The power spectral analysis of this sequence revealed periodicities of 2–3, 7–8, 21 and 80 years, hot summers being alternated with cool ones. The hot summer intervals were 1730–80, 1820–70 and 1920–50 AD, whereas the cool ones were pre-1730, 1780–20 and 1870–1990.

According to Zhang De'er (1991), an analysis of historical literature for the last 2000 years shows no linear trend for moisture variation but rather repeated alternation of dry with humid or rainless with rainy intervals. The author used the frequency of large-scale storms, plus flood records of the Yellow River, to calculate annual moisture variations for the last 1000 years. He also compared these data with the dust fall frequencies and with historical data from all over eastern China, including the *Qingyulu* (see above) for NanJing, Suzhou and Hangzhou (1720–1800), which contained reports of the Meiyu (plum rain). On the whole, Zhang De'er came to the conclusion that the climate of the past 1000 years displayed a quasi-periodic variation that was related to the quasi-periodic features in atmospheric circulation. The twentieth century was located at a position in the warmer interval with normal precipitation over the last 1000 years period.

Wu Xiangding and Zhan Xuzhi (1991) extracted proxy-data of climatic change from tree ring width in three main regions of China.

1. The Xizhang Plateau of Tibet. In this region, distinct fluctuations of temperature during the first 2000 years were recorded. The climate was cold in the early first century AD. Then it became warmer and was 1 °C warmer during the second and third centuries. It maintained a longer colder period from the late third century to the fifth century. Generally it was about 1 °C lower than at present. From the sixth century to the late twelfth century, the increase in temperature was more remarkable on the plateau. During the thirteenth century, there was a distinct cold period. During the seventeenth century the climate was the coldest of the past 1000 years. It has been getting warmer since the mid nineteenth century. For the last 340 years, the variations of drought and floods in the Tibetan plateau, derived from tree ring series, show that the major wet period stretched from the mid seventeenth century through the early eighteenth century. Another wet period was from the end of the nineteenth century to the 1930s. The plateau has experienced a drying tendency during the last 30 years. The authors also remark that, according to the features of climate change in Tibet during historical times, it can be

deduced that the plateau is an area more sensitive to climate change than other areas on the same latitude.

2. The Hengduan mountains. In this area, a dry cold pattern was prevalent during the first half of the seventeenth century AD. In the second half of the same century, the climate became warm and wet. In the nineteenth century, the number of both cold and wet years increased. The dry warm pattern has been increasing since the beginning of the twentieth century. On the whole, the variations in air temperature in this region are smaller than in Tibet and east China.

3. Northwest China. The general trend of temperature fluctuations over the last several hundred years has been similar to that of the eastern part of China.

Xu Guochang and Yao Hui (1990), from the Institute of Desert Research at Lanzhou in the arid northwestern region of China, suggested three modes of division of the Holocene according to different time scales. They based their division on data from glaciers, geology, lakes, tree rings, ice cores, historical data, etc.

1. On a general time scale, they suggest dividing the Holocene into three parts:
    Early Holocene, 10 ka to 7.5 ka BP: cold and dry;
    Middle Holocene, 7.5 ka to 3.5 ka BP: warm and wet;
    Late Holocene, 3.5 ka BP to the present: cold and dry;
2. On the 1000 year scale:
    8.7 ka to 7.5 ka BP: first cold stage;
    7.5 ka to 5.9 ka BP: first warm stage;
    5.9 to 5.4 ka BP: second cold stage;
    5.4 ka to 4.2 ka BP: second warm stage;
    4.2 ka to 3.8 ka BP: third cold stage;
    3.8 ka to 3 ka BP: third warm stage;
    3 ka to 2.7 ka BP: fourth cold stage;
    2.7 ka to 2 ka BP: fourth warm stage;
    2 ka to 1.4 ka BP: fifth cold stage;
    1.4 ka to 1 ka BP: fifth warm stage;
    1 ka BP to the present: sixth cold stage.

For most of the stages, the cold corresponded with dry and the warm with wet.

3. On a 100-year scale for the last 1000 years
    main cold spells appeared during the decades following 1100, 1310, 1470, 1680 and 1820 AD
    main warm spells were in the decades following 1200, 1390, 1570, 1780 and 1880
    more recent changes in precipitation for most parts of the arid and semi-arid regions of China are mainly a rainy and cold stage, *c.* 1800 AD, and a warm and dry period, with an increase in drought frequency, in the twentieth century.

Wan Jingtai *et al.* (1991) reported on natural hazards in the Gansu Province of China. Most of this district lies in the semi-arid and sub-humid zone, within the western margins of the monsoon region of China. The summer wind pattern has a very strong influence on the climate of southern Gansu, the variations in precipitation from year to year being very large. As a result, droughts are frequent and very severe. Climatic records since the time of the Han Dynasty (about 2000 years ago) to the present show that some 315 droughts have occurred, of which 43 were very severe. These authors claim a cyclic periodicity of 121, 64, 32, 15, 8–9, and 2–3 years. The cycles have become shorter and have smaller amplitudes starting about 200 years ago, and it seems clear that the climate of eastern Gansu has become progressively drier.

Fan Jianhua and Shi Yafeng (1990) investigated the impact of climatic changes on the hydrological regime of the big inland lake Qinghai, which may be taken as representative of the impact of climatic changes on the hydrological cycle in northwestern China and central Asia. Since the 1960s the climate in the lake basin has had a tendency to warm up and become drier. During this period, the number of dry years accounted for 60% of the total years, and from the early 1960s to the early 1980s, the annual temperature increased by 0.3 °C. Because of the warming and drying, the lake level has continuously decreased.

Bradley *et al.* (1987) studied the secular fluctuations of temperature over northern hemisphere land areas and mainland China from the mid nineteenth century. Records for the two regions showed similar trends: temperatures increased from the late nineteenth century to a maximum in the 1940s, followed by a cooling trend that reversed over the final 10–15 years. Extremely sharp falls in temperature, particularly in autumn months, occurred after several major volcanic eruptions. High temperatures were sometimes associated with major El Niño years. When the two occurred more or less simultaneously, their influence was minimized.

By comparing proxy-data, Lough *et al.* (1987) investigated the relationship between the climates of China and North America over the past four centuries. They found that drought in China appeared to be directly and significantly correlated with a weakening of the north Pacific sub-tropical anticyclone (NPSA). This was associated with above average precipitation in the central and southern USA. Floods in China were associated with intensification and expansion of the NPSA anticyclone system, which also caused a decrease of precipitation in the USA. In central Asia, the Upper Holocene climatic variations were investigated by tracing the movements of the glaciers of the Tienshan mountains. Three major glacial advances were traced, at *c.* 2100–1700, 1300–1100 and 700–150 lichenometric years ago (Savoskul and Solomina, 1996).

3.1.4 Korea

Kim and Choi (1987) reported on a preliminary study of long-term variations of unusual climatic phenomena during the past 1000 years in Korea. They found the following eight periods.

1. 1381–1420: cold period (winter warm, summer cold);
2. 1421–1520: warm period (winter and summer warm);
3. 1521–1630: first period of Little Ice Age (winter and summer cold);
4. 1631–1740: second period of Little Ice Age (winter cold, summer very cold);
5. 1741–1840: third period of Little Ice Age (winter and summer cold);
6. 1841–1880: warm period (winter and summer warm);
7. 1881–1930: cold period (winter and summer cold);
8. 1931–present: warm period (winter and summer warm).

They also observed that cold winters in the early nineteenth century corresponded to periods of weak sun activity.

Cool weather phenomena in summer seasons in the Korean Eastern Sea region were explained by an intensification of the Okhotsk High in southern or southwestern directions in summer seasons.

Liu Changming and Fu Goubin (1996) analyzed the records of temperatures, river flow, levels and areas of lakes and movement of glaciers during the twentieth century. They conclude that, although the monsoon regime determines the hydrological cycle, the nature, range and scale of the impact of the monsoons differ regionally. They found that there was a period of prolonged low flow in North China during the last century, especially during the warmer period from the 1920s to the 1940s and during the 1980s. In general, the number, area and volume of lakes have been reducing, while glaciers tended to decline.

3.1.5 Conclusions

As can be seen from the data summarized above, the climate changes during the Holocene in China were many and differed along the dimensions of space and time. The differences are a consequence of the vast area covered by China, various parts of which are influenced by different climate regimes and by their interplay. While in general it can be concluded that the part of China, mainly the southeast, which is mainly dominated by the monsoon regime will get more precipitation, and thus floods, once global temperatures rise, its northwestern parts will see less precipitation and may suffer from droughts. The spatial extension and magnitude of these changes depend on the interplay between the climate regimes of the Pacific and Indian oceans on the one hand and the continental mass of central and northern Asia on the other; special geographical features of each province of China will also have a modulating effect.

3.2 JAPAN

3.2.1 Contemporary climate

The climate of contemporary Japan is determined by its position to the southwest of the Siberian sub-continent and on the northeastern fringe of the Pacific Ocean. Consequently, a large cold air mass is conveyed from Siberia during the winter, forming the northwestern monsoon, while during the summer, the southeastern monsoon transports warm moist air masses originating in the sub-tropical high of the northern Pacific (Maejima, 1980).

Consequently, during the various seasons, Japan's weather is determined by two main pressure systems, cyclonic and anticyclonic, each divided into three subsystems.

1. The cyclonic systems:
    extratropical cyclone subsystem;
    frontal subsystem;
    typhoon subsystem.
2. The anticyclonic system:
    winter monsoonal subsystem;
    migratory subsystem;
    summer monsoonal subsystem.

The dominance of these systems and subsystems is determined according to the seasons and is decided by the interplay of the upper westerlies waves, the jet streams, and the blocking Okhotsk high – a subsidiary of the Siberian high.

In winter, the synoptic surface pattern is decided by the Siberian high and Aleutian low. The Siberian high is very cold, not only because of its continental character and northern location but also because the Tibetan plateau blocks the mitigating influence of the Indian Ocean. The high pressure over Siberia instigates the northwesterly monsoon system, moving cold dry air masses over the Sea of Japan. During their passage over this sea, they absorb heat and moisture. Reaching the coast of eastern Japan, orographic lifting, combined with convective conditions, cause the unstable air masses to precipitate heavy snow. These air masses dry up and then descend over the leeward side of the mountainous axis range of Japan. Thus, during winter, the side of Japan facing the Pacific is characterized by windy but sunny weather. This starts at the end of September and continues until the end of February.

During the summer, two main systems prevail, the Baiu type and the midsummer type, which are functions of seasonal displacement of the upper westerlies and of the Northern Pacific high. Circulation over Japan during the summer starts with the shift of the sub-tropical jet stream from the south of the Tibetan plateau to its north; it then splits into two jet streams over Japan, with the Okhotsk high as the blocking factor between the two streams. Thus, the beginning of the Baiu season in Japan coincides with the starting of the Indian monsoon, which is also related to this shift.

In the area of the Okhotsk high, cool humid air masses develop, which bring northeasterly winds and cool and cloudy weather to the Pacific coast of northern Japan. At the same time, a front is formed along the convergence zone between the Okhotsk Sea air mass and the warm Ogasawara air mass, originating in the Northern Pacific high. This brings with it rainy weather, which travels northward and southward over the southwestern and central parts of Japan.

In the latter part of the Baiu season, a hot and moist equatorial air mass invades western Japan and causes torrential rain. This lasts from early June to mid-July. During mid-summer, the southern jet stream shifts from the Pacific coast of central Japan to northern Hokkaido. This causes convective instability in the hot and moist air masses over the Northern Pacific high, bringing the southeastern monsoon to Japan, characterized by high temperature and humidity. This weather prevails to the end of August. From this time to the middle or end of October, the Shuin rainy season dominates. It starts with the migration of the westerlies southward, while the Northern Pacific high becomes weak. During this season, most of the catastrophic typhoons in the region of Japan occur (Maejima, 1980)

3.2.2 The climate of Japan during the Last Glacial and Post Glacial Periods

Low temperatures with some fluctuations in moisture and humidity characterized the climate of Japan during the Last Glacial and Post Glacial Periods. Field surveys of river terraces and hill slopes, by Ono and Hirakawa (1975) and Oguchi (1988), indicated that the climate during isotope stage 4 (i.e., *c.* 75 ka to 58 ka BP (Shackleton and Opdyke, 1973)) was wetter than that during stage 2 (i.e., *c.* 24 ka to 14 ka BP (Shackleton and Opdyke, 1973)). Markedly dry conditions during stage 2 were confirmed by palynological research. In particular, the frequency of storms induced by the Baiu front and typhoons was considerably reduced because of the southward shift of frontal zones (Suzuki, 1979). This climatic condition accounted for inactive fluvial processes and reduced hill-slope erosion.

During the late glacial, heavy storms, caused by typhoons and the polar front, significantly increased in response to climatic warming. This led to active hill-slope erosion and fluvial deposition at piedmont areas (Hatano, 1979; Oguchi, 1988, 1997). Snowfall in winter also increased along the region facing the Sea of Japan, reflecting more moisture supply from the ocean.

Although temperature and moisture in and around Japan tended to increase during the Pleistocene–Holocene transition, climatic cooling during the Younger Dryas occurred, as it did in Europe (Aoki, 1994; Koizumi and Aoyagi, 1993).

3.2.3 Climate during the Holocene

3.2.3.a PALYNOLOGICAL TIME SERIES

The climate variability during the Holocene was investigated using palynological time series and archaeological and historical records. Sakaguchi (1982) described climatic variability during the last 7600 years based on a continuous section of peat samples taken from the wall of a pit in the raised bog of the Ozheghara moor, located near the head of the Tadami river, which flows into the Sea of Japan, 150 km north of Tokyo (site P73). The age was determined by the corrected ^{14}C ages of the peat taken at 50 cm intervals and, additionally, by the dates of the tephra beds, which are intercalated in the peat layers. The age of the latter is known either through ancient manuscripts or archaeological findings. For ages younger than 2.45 ka BP, the non-corrected ^{14}C readings were used because the difference between the corrected and the non-corrected readings are small (the corrected being younger by 20–80 years). At 450 cm deep, the non-corrected age is 7.03 ka BP and is slightly over the correction limit given by the tree ring of *Pinus aristata*, dated at 4760 BC. Sakaguchi then suggested a correction factor of +600 years. Thus, the corrected age of 7.03 ka BP should be 7.6 ka BP. The Ozegahra moor lies 1400 m above sea level. The trees in this zone are mainly deciduous. Conifer trees increase in number above 1550 m of which some species cannot withstand snow cover, so their decline indicates cold temperatures and snow cover.

Sakaguchi (1982) took the percentage of *Pinus* as the characteristic for climate change as it reflects the flourishing or decline of the sub-alpine conifer forest zone, which, in turn, reflects the downward or upward shifting of the vegetation zones (i.e., the fall or rise in air temperatures). In a later paper (Sakaguchi, 1983), he generalized the palynological diagram of site P73 and compared it with paleo-climatic diagrams from all over the world. A correlation with Japanese culture was also performed. The following sequence was obtained:

1. 5660–4500 BC: a warm stage (Early Jomon warmest stage);
2. 4500–2450 BC: warm;
3. 2450–2270 BC: cold (Middle/Late Jomon cold stage);
4. 2270–2090 BC: warm (first Late Jomon warm stage);
5. 2090–1460 BC: warm;
6. 1460–1250 BC: warm (second late Jomon warm stage);
7. 1250–870 BC: warm/cold (transitional stage);
8. 870–400 BC: cold (latest Jomon cold stage);
9. 400–20 AD: warm (Yayoi warm stage);
10. 20–240 AD: warm (cold transitional stage);
11. 240–730 AD: cold (Kofun cold stage);
12. 730–1300 AD: warm (Nara–Heian–Kamakura warm stage);
13. 1300–1900 AD: cold (Little Ice Age).

Sakaguchi (1983) regarded the period between 870 and 400 BC (his dates 866–398 BC) as the first long-term cold stage since 7.6 ka BP. During this period, according to various reports quoted by the author, the water level of Lake Ogawara sank to below the present level (*c.* 3 ka BP) and a regression phase occurred in the Tsugaru plain, evidenced by peat layers and buried trees (reaching a peak *c.* 2.5 ka BP).

Another cold stage (following the Yayoi warm stage, which began *c.* 400 BC), which Sakaguchi (1983) calls the Kofun cold stage, started *c.* 2 ka BP (17 AD) and lasted until 730 AD with cold climaxes appearing around 270 AD and 510 AD. A warm episode occurred *c.* 390 AD. The author quoted a few sources that present geomorphologic evidence showing that the sea was 2–3 m below that of the present during the cold stages. Sakaguchi (1983) came to very interesting conclusions regarding the impact of the climate of the Kofun cold stage, the intensity of which, he claimed, exceeded that of the Little Ice Age. This period was called Kofun because huge earthen grave mounds, called kofuns, were constructed for the ruling class during this period. These, with other historical evidence, tell of a highly organized centrally controlled society. The cold climate was not favorable for a rich rice harvest (irrigated rice cultivation in Japan is considered to have begun in the latest stage of the Jomon period in North Kyushu and then to have spread to eastern Japan in the Yayoi period). Moreover, the period was characterized by frequent floods. These conditions led to the development of elaborate agricultural techniques, especially of a type of irrigation of the paddy fields. The culmination of the Kofun culture corresponded with the warmer interval *c.* 400 AD.

Later, Sakaguchi (1989) revised his diagram (Fig. 3.3). In his paper of 1989, Sakaguchi also described a pollen section of a core 940 cm long taken from the Oshima peninsula, in the southern part of Hokkaido Island, northern Japan (site K87). The age of the bottom layer is 23.7 ka BP. On the basis of ratios between the different pollens, the author divided the coldness levels into a scale ranging from 1 (warmest) to 5 (coldest), when 4 to 5 was the range during the last glacial period, 4–2 from 13 ka to 8.8 ka BP, with short cold periods of grade 4 from 10.2 ka to 10 ka and from 9.4 ka to 8.8 ka BP. The warm climate continued up to the present, with an intercalation of a cooler climate from 1.55 ka to 0.89 ka BP.

Generalizing Sakaguchi's (1989) pollen diagram of site K87 by calculating a running average diagram for the abundance of *Betula* (birch), *Alnus* (Alder) and *Quercus* (oak) pollen (Fig. 3.4), one can see that, in the lower part of the Holocene (and the same goes for the pre-Holocene period, which is not shown on the present diagram), one can observe an abundance of *Betula* pollen, while that of *Quercus* is very low. In other words, the *Betula* and *Quercus* curves change inversely to each other. For this period, it can be concluded that the abundance of *Betula* and scarcity of *Quercus* were mainly a function of a colder period and vice versa. During the period between 7 ka and 4 ka BP, however, this relationship cannot be observed, as both *Betula* and *Quercus* were relatively rather low. During the same period, the non-arboreal pollen was

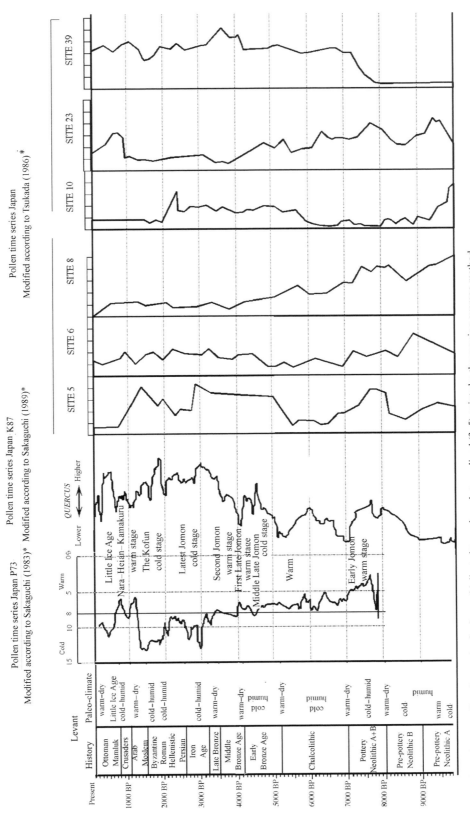

Fig. 3.3. Holocene pollen time series for Japan. *Adjusted to scale and streamlined (3–5) points by the running average method.

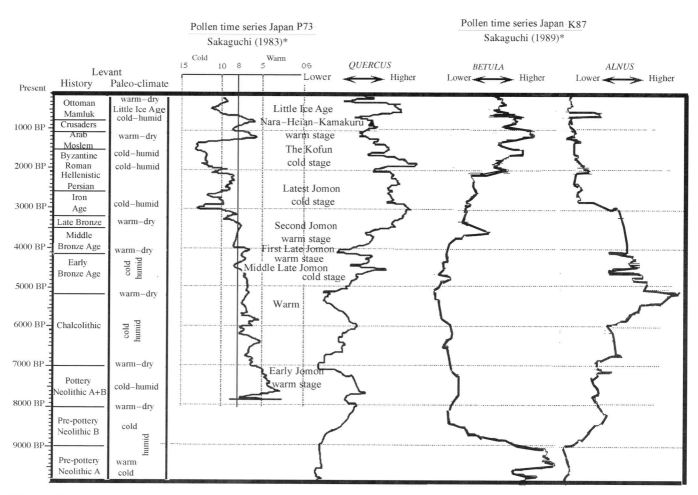

Fig. 3.4. Holocene pollen time series for Japan generalized from the P73 site data in Sakaguchi (1983) and K87 site data in Sakaguchi (1987). (See text for details.) *Adjusted to scale and streamlined (3–5) points by the running average method.

relatively abundant. The fact that *Pinus* was also at a minimum may suggest that, during this period, the environmental situation, although relatively warm, was for other reasons not favorable for *Pinus* or *Quercus*.

Consequently, when trying to correlate the diagrams of sites P73 and K87 (Fig. 3.3), and taking into account the different climatic and, thus floral, environments, it is suggested that we look at what happened during the extreme and sufficiently lengthy climatic phases, which can be assumed to have had some impact on the floral environment all over Japan. (One should also take into account variations in the dating system between the two sites: the dates of site P73 were carried out at 50 cm intervals and these age measurements were corrected, while at site K87, the density of dating was not as precise, with no report of age correction.) With these considerations in mind, one can nevertheless see that one of the main long cold periods (i.e., the

Kofun cold stage) caused an increase in the pollen of *Betula* and a reduction in *Quercus*. Later, the Nara–Heian–Kamakura warm stage was characterized by more *Quercus* pollen. Yet, Sakaguchi (1989, p. 1) came to the conclusion that "K87 possibly belongs to a different climatic region from P73, at least after 8.8 ka BP". I would reinterpret this conclusion by suggesting that, while the geographical environments of the two sites went through the same climatic changes, the process had distinct effects during the extreme and lengthy changes in climate but less distinct effects on the flora when, climatologically speaking, the changes were less distinct.

Miyoshi and Yano (1986) investigated the pollen of a core drilled in a moor in the Chugoku mountains of western Japan. The section was divided into zones and sub-zones as follows:

zone A1, 20 ka to 14 ka BP: sub-Arctic and dry;
zone A2, 14 ka to 11.5 ka BP: sub-Arctic with cool–temperate/ dry–wet;
zone B, 11.5 ka to 8 ka BP: cool temperate/wet;
zone C1, 8 ka to 4 ka BP: temperate/wetter;

zone C2, 4 ka to 1.7 ka BP: cool temperate/wetter;
zone D, 1.7 ka BP to present: cool temperate/wetter.

In his study of the vegetation of Japan during the last 20,000 years, Tsukada (1986) used data from 40 separate locations. Figure 3.3 presents five out of his twenty-four *Quercus* pollen time series together with the climate change profiles based on the pollen diagrams presented by Sakaguchi (1983). As can be seen, a clear reduction in *Quercus* is evident at sites 5 and 6, which are at a latitude of *c.* 33° N, and at site 39 at a latitude of *c.* 44° N. However, one should consider that *Quercus* responds negatively mainly to a cold setback in temperate regions, while in a warmer climate a positive response depends on a number of factors: plenty of humidity and environmental factors, such as other competing floral societies.

As discussed above, the sequence of climatic changes suggested by Sakaguchi, (1983, 1989) based on the palynological section of site P73 is the most detailed and could be correlated, in terms of major changes, with other time series in Japan. Therefore it is suggested that Sakaguchi's division be regarded as the base curve for Japan. Unfortunately, this section does not extend prior to *c.* 8 ka BP.

3.2.3.b GEOMORPHOLOGICAL OBSERVATIONS

Being situated in a region of pronounced tectonic activity, Japan is also a region where the movement of sea terraces, owing to tectonic activity, has to be taken into consideration. Japanese scientists examined the sea shores of Japan and concluded that sea transgressions were responsible for high terraces dating from 6.5 ka to 5 ka BP (peaking between *c.* 6.5 ka and 6 ka BP). This post-glacial transgression is called the Omon transgression. Many deposits, known as middens, were left by this transgression. Between 3 ka and 2 ka BP, the so-called Yayoi regression occurred. This was inferred from the occurrence of shallow buried valleys and the development of coastal dunes. Another regression, called the Middle Jomon minor regression took place *c.* 5 ka to 4 ka BP (Yonekura and Ota, 1986). Pirazzoli and Delibrias (1983), investigating Late Holocene and recent sea-level changes and crustal movement in Kume Island, claimed that the maximum Holocene mean sea level was probably reached *c.* 4 ka BP, while a gradual fall occurred between at least 3 ka and 1.2 ka BP. Sakaguchi (1983) stated that the Middle/Late Jomon stage (2450–2270 BC) was characterized by a fall in seawater temperature, as evidenced by the ^{18}O composition of *Meterix lamarcki* (Sakaguchi, 1983, refering to data from Chinzei *et al.,* who investigated the paleo-oceanography of the Pacific along the east coast of Honshu) which indicated a weak but rather long cold episode from 5 ka to 4 ka BP. Sakaguchi (1983) also quoted Ota *et al.,* who investigated the Holocene sea levels in Japan and found a minor regression between 5 ka and 4 ka BP, and Matsushima, who described a decrease in the number of warm

water/shallow water molluscan assemblages in the deposits in the southern Kanto region from 5 ka and their disappearance *c.* 4.5 ka BP.

However, high sea levels were recorded at 4 ka and 3.5 ka BP. Sakaguchi (1983) correlated the climatic changes at site P73 with the number of other archaeological sites and shell middens at high levels and found that these both occurred from *c.* 4.3 BP to *c.* 3.3 BP.

Yonekura and Ota (1986) cite Furukawa, who dated shells in middens to *c.* 4 ka BP in the Nobi plain of central Honshu, and Hirai, who observed that Lake Ogawara was at a high level between 5 ka and 4 ka BP. As this lake is connected to the sea, it should be considered as a high sea-level mark.

Sakaguchi (1982) suggested that the period around the twelfth and thirteenth centuries was colder and humid, based on finding a layer of detritus. However, in a revised paleo-temperature curve (Sakaguchi, 1989), these periods were interpreted as being warmer. During the warm period between 730 and 1300 AD, called by Sakaguchi (1983) the Nara–Heian–Kamakura warm stage, the sea level was higher in the area of Fukuyama city (Sakaguchi, 1983; citing Kuwashiro (1965)).

During the Little Ice Age, which started in Japan *c.* 1300 AD, there was a warm episode at ca 1840 and an extremely cold period around 1890. During this last episode, in which storms were abundant, rice harvests were low, leading to starvation, panic and rebellions. Sakaguchi (1983) concluded that, overall, the Little Ice Age was relatively moist in Japan.

Based on historical and other relevant data from 1650 to 1983 AD, Takahashi (1987) analyzed the long-term variation in storm damage in Japan. He found a 70 year cycle, which tended to increase in frequency when the climate was warm. The analysis indicated the considerable influence of volcanic activity on climatic change and, hence, on storm damage over a time scale of several decades.

Mikami (1987) reconstructed the climate of Japan from 1781 to 1790 AD, in comparison with that of China, in order to derive data from processing Chinese historical documents. The Japanese climate in the late eighteenth century was estimated to have been very cool and wet, as severe famines caused by crop failure were frequently recorded. From 1782 to 1787, Japan experienced the most severe famine in its history, causing a decrease in its population by approximately one million. The statistical examination of the relationships between Japan and China show simultaneity of rainfall in Japan and central China, with opposite indications in north and south China. These are considered to be linked to global circulation patterns. For example, in 1978, when it was extremely dry and hot in Japan and middle China, the axis of the sub-tropical high (North Pacific high) was located to the north. Such a situation most probably also occurred in 1785. In 1980, when it was exceedingly wet and cool in Japan and middle China, the axis of this high was shifted to the south and the polar frontal zone extended from middle China to Japan. Such a situation most probably also existed in 1783.

Oguchi (1997) suggested that since the transition from the Pleistocene to the Holocene, which led to the northward shift of frontal zones, typhoons started to hit Japan more frequently, causing frequent and heavy rainstorms and leading to extensive erosion by running water. The resultant incised channels were cut into regolith as well as consolidated bedrock because of the strong erosive force during storms. Channel formation is still active in most basins. Accordingly, erosion rates and sediment yields subsequent to rapid channel expansion in these areas will increase significantly if future climatic changes lead to increased storms.

3.2.3.c GENERAL CONCLUSIONS

From the research studies surveyed in this chapter, it can be concluded that during the last glacial period, between 20 ka and 18 ka BP, the climate all over Japan was drier than today. However, the degree of difference between then and now varied from north to south. It was very dry in Hokkaido as well as in Honshu. This was so because the Siberian high-pressure zone was very strong in winter, causing a strong winter monsoon. Around 10 ka BP, while it was still colder than today, the northern regions of Japan were drier because the strong Siberian high still prevailed, while the central region was wetter. In the northern Japanese Alps, for instance, strong rains and floods occurred. Around 6 ka BP, it was very warm all over Japan. The island of Hokkaido was dry, while central Japan was rather wet. The dry Hokkaido conditions may have been caused by the weak winter monsoon, which resulted in decreasing snow accumulation. After c. 4 ka BP towards 3 ka BP, the climate became cooler, reaching a cold maximum around 2.5 ka BP. During this period, most of Hokkaido was dry

(except in the deep south), while most of Japan was wet. The formation of sand dunes in Kyushu shows that this region was also dry (Yoshino and Urushibara, 1979).

As was discussed by Issar (1995b, pp. 55–65), a good correlation was found between China and the Levant and, therefore, it is only logical that such a correlation should also include Japan, as shown in Figure 3.4.

The most outstanding correlation is between the Kofun cold stage in Japan and the Roman–Byzantine cold stage in the Levant, as well as between the subsequent Nara–Heian–Kamakura warm stage in Japan and the Moslem–Arab warm period in the Levant.

From Sakaguchi's curves for the P73 and K87 sites (Sakaguchi, 1983, 1989), it can be concluded that the latest Jomon cold Stage, c. 3 ka BP, was actually composed of three sub-stages, at c. 3.3 ka, 3 ka and 2.6 ka BP, which, in general, correlate with the Iron Age cold period of the Levant, which started c. 3.2 ka and continued until c. 2.7 ka BP (Issar, 1995a, b). Suzuki (1979) maintained that the most extreme spell occurred at 3.5 ka BP.

The First Late Jomon and Second Jomon warm stages can be correlated with the MB and Late Bronze warm periods of the Levant, while the Middle Late Jomon cold stage can most probably be correlated with the EB cold period of the Levant.

An abrupt bend in Sakaguchi's curves towards a colder trend, which was not given a special name by him, correlates with the Lower and Middle Chalcolithic cold period of the Levant, if one agrees to push this phase a few hundred years back on the time dimension. As the accuracy of the age determinations of the climatic changes during the lower part of the Holocene is in the order of magnitude of a few centuries, (because of the difference between corrected and non-corrected ^{14}C dates), such a correlation looks feasible. The same can be said with regard to the Middle Neolithic cold period of the Levant and the cold period of the Base Neolithic.

4 Climate changes during the Holocene in Africa

4.1 EGYPT

4.1.1 Contemporary climate and hydrological regime of the Nile

From the climatological point of view, Egypt is the northeastern edge of the Sahara belt (Fig. 4.1) and its extreme aridity is caused by the descent of tropical air masses, which causes them to become hot and dry. Along the Mediterranean coast, a narrow strip of its land is influenced during winter by the cyclonic system of the westerlies. The annual average precipitation over most of Egypt is less than 50 mm and only along the coast does it reach about 100 mm. Yet Egypt, since ancient times, has been inhabited by an agricultural society because of the water brought by the Nile (Fig. 4.1). This river is the longest in the world, measuring about 6600 km from its headwaters in Rwanda to the northern edge of its delta. It has a relatively high constancy of flow and of periodicity of its fluctuations. There are several reasons for this. The Nile is fed by precipitation falling on the subequatorial countries of east Africa, the source of this moisture being the southwesterly air streams from the Indian Ocean and the equator during the northern hemisphere summer. These summer rains affect mainly the Blue Nile and the Atabara rivers, which originate in the north and central highlands of Ethiopia. The resultant floods start in June and reach their peak in August. The White Nile is fed by rains falling on equatorial Uganda and southern Ethiopia, where there are two rainy seasons, thus providing more evenly distributed runoff. Another very important storage and regulating factor is the gigantic swamps of the Sudd into which the White Nile flows; here the suspended silt of the river is deposited and the supply of water is regulated. In terms of relative contributions, the Blue Nile contributes about 60% of the average annual water supply, the White Nile about 30% and the Atabara about 10%.

4.1.2 Climate during the Upper Pleistocene and Holocene

Between 20 ka and 12.5 ka BP, the Nile was a highly seasonal, braided river. Towards the end of this period, from 14.5 ka to 12.5 ka BP, the amount of water flowing through the Nile began to increase and several high sub-stages can be recognized, as well as moderate activity in the wadis (Adamson et al., 1980; Butzer, 1980). Around 12.5 ka BP, there was a huge increase in the discharge of the Nile, partly caused by overflowing east African lakes, which expanded the river's catchment area. The overflow from Lake Victoria and other lakes, as well as higher rainfall in Ethiopia, sent extraordinary floods down the main Nile. This marked a revolutionary change from an ephemeral to a continuous flow, with a superimposed flood peak. As a result, the main Nile and its tributaries formed more stable channels of higher sinuosity, from which suspended Ethiopian silt and clay was deposited on the floodplains (Adamson et al., 1980). The "wild" Nile regime, characterized by floodplain sedimentation, lasted until c. 11.5 ka BP and was followed by a period of strong dissection and down-cutting in the Nile and associated wadis of at least 20 m (Butzer, 1980; Fairbridge, 1962).

Later on, two periods of high Nile levels and aggradation can be recognized in Egypt: 11.2 ka to 7.7 ka BP and 7.3 ka to 6 ka BP. These two periods are separated by a regression (Butzer, 1980), apparently caused by a more arid climate in the entire region of northern Africa, as recognized by falling lake levels c. 7.5 ka BP (Street-Perrot et al., 1985).

Butzer (1980) summarized the main features of upper Holocene history of the Nile valley in Egypt and lower Nubia.

1. 3000 to 2800 BC. Flood levels declined significantly, representing an overall reduction in volume of 25–30%. The concomitant down-cutting appears to have initiated the modern flood plain downstream of Wadi Halfa (Bell, 1971).
2. 2250 to 1950 BC. A period of catastrophically low floods.
3. 1950 to 1840 BC. Improved floods.
4. 1840 to 1770 BC. Excessive floods are documented, reoccurring every 2 to 5 years, with peak discharge three times that of the ten greatest floods of the nineteenth century AD (Bell, 1971).
5. 1770 to 1180 BC. Average levels remained high.
6. 1180 to 1130 BC. Strong decline in levels.

Fig. 4.1. Map of Africa.

7.   1130 BC to 600 AD. "Normal" levels.
8.   600 to 1000 AD. Generally high levels.
9.   1000 AD to present. "Normal" levels.

The present level of the Nile was reached about 5 ka BP. Archaeologically dated high-water marks on Egyptian temples and associated Christian constructions near Wadi Halfa suggest that the last important high Nile periods were *c.* 500 AD and 800 AD (Fairbridge, 1962).

Flohn and Nicholson (1980) observed two separate wet periods: the first, *c.* 9.5 ka BP, coinciding with the warmest Holocene period in southern latitudes and the second, *c.* 6 ka BP, coinciding with the thermal maximum in Europe (Atlantic). The very dry periods in the Sahara were between 7 ka and 6 ka BP and from 1 ka BP to the present. According to Nicholson (1980), the level of the Nile during the Roman period was low.

Fairbridge (1984) analyzed the flood levels of the Nile, by the Fourier and maximum entropy statistical analysis. He found cyclic fluctuations, which he attributed to various factors: first, an 18.6 year cycle that is connected to the lunar nodal notations; second, a 78 year cycle that, in his opinion, is a solar cycle. He found only a weak reflection of the 11–22 year solar cycle.

Sneh *et al.* (1986) reconstructed the evolution of the northeastern corner of the Nile delta by investigating a column of sediments obtained from a 48 m drill hole divided into five units. They found that only a minor part of the sediment derived from the Nile in the early Holocene (unit 1, 42–48 m), the main part being marine littoral sands. The shoreline at that time was more than 20 km south of that of the present. The ^{14}C age of the solitary corals found in this layer is 8480 ± 280 BP. The minor influence of the Nile, even though it was at high levels at this time, may be explained by the fact that it had not yet developed any eastern branch. Overlying the marine sands was a layer of black clays (unit 2, 42–30.5 m), which is evidence of an influx of sediments derived from land soils, yet no evidence of the typically tropical material supplied by the Nile was to be found. The lower part of this section was of marine character, containing foraminifers, ostracodes and molluscs. In the upper part of the unit, the influence of the Nile was more pronounced, with freshwater diatoms, but there was still a relative lack of tropical pollen, showing that the outlet of the Nile was still in the west and, therefore, no sediments could have been brought by the sea currents which move from the west to the east. What pollen there was in the core was typical of desert vegetation. The nearby seashore retreated away to the south of the borehole site. Sneh *et al.* (1986) related this terrigenous extension to high floods in the upper part of the Nile and correlated it with the upper sapropel (i.e., deposits rich in organic material), which reflects a major influx of freshwater into the Mediterranean sea (Luz, 1979; Rossignol-Strick *et al.*, 1982).

In unit 3 (30.5–17.5 m), there were layers of clayey silt, also deposited in a marine environment, the lower part of which showed a marked increase in tropical fern spores and pollen of freshwater plants, while the amount of desert plants sharply decreased. In the upper part of unit 3, although the deposition of silts continued, there was a marked change in the flora to few tropical elements, less freshwater pollen but a relative abundance of pollen of sabkha and saline marsh vegetation. Sneh *et al.* (1986) suggested that this was a regressional phase.

In unit 4 (17.5–8.5 m), the sediments were of a typically pro-delta character, with alternating silt and clay laminae. There was also an increase in the tropical elements. Salinity, determined by the "sieve pore" shapes of the ostracoda, decreased. The pollen spectra suggest that a warm and dry climate dominated lower Egypt.

Unit 5 comprised the uppermost 8.5 m of the section. A ^{14}C age of 2890 ± 220 years was found at the base of this unit, indicating that it was deposited during the last 3000 years. The Pelusian channel, which is the most eastern tributary of the Nile, determined the character of the layers. The sediments were mainly those of coarse silts and sands. The pollen assemblage showed an increase in tropical spores and a relative decrease of delta plain pollen. There was also some increase in Sabkha vegetation. The fauna suggested deposition in a saline lagoonar environment. Consequently, this section represents the stage of the seaward build-up of the eastern part of the delta as a function of the outlet of the Pelusian channel.

Foucault and Stanley (1989) also investigated the Quaternary paleo-climatic oscillations in east Africa as recorded by the heavy minerals in the Nile delta. The minerals are supplied by the White Nile tributary, which drains the tropical region and which contributes about a third of the Nile's discharge. It has a relatively low sediment load, because of the dumping effect in the Sudd swamps. As the White Nile comes from a region constructed of metamorphic rocks, the proportion of amphiboles in its sediments is high. While the Blue Nile supplies more than half of the Nile discharge, it contributes about three quarters of its sediment supply. As it drains the volcanic Ethiopian highlands, the proportion of pyroxenes to amphiboles is high. The Atabara contributes one quarter of the Nile's sediments, which are also rich in amphiboles. The percentage ratio of amphiboles to amphiboles plus pyroxenes, which is called the amphibole index, was investigated in core holes in the Nile delta in order to elucidate variations in the paleo-climates that had affected its tributaries. At present, there is a high pyroxene load in the summer floods because of the high rainfall on the rather arid Ethiopian highlands. Foucault and Stanley reasoned

that a more humid climate over this region would increase the vegetation cover, and thus reduce the sediment load, as well as the quantity of pyroxenes derived from this region. They found that high amphibole ratios *c.* 40 ka to 20 ka BP corresponded to periods with high levels of water in the lakes east of the Ethiopian plateau as well as in lakes in the Ethiopian rift. A low amphibole ratio *c.* 20 ka to 12 ka–10 ka BP corresponded to low lake levels.

Stanley and Warne (1993) analyzed mineralogical, textural, faunal and floral content and trace element geochemistry of the sequence of layers found in more than 80 core holes in the northern part of the Nile delta. They found three distinct lithofacies sequences. The lower sequence, dated from *c.* 35 ka to 12 ka BP, was one of non-marine alluvial sandy deposits. Interleaved with these sands were variegated mud layers. These layers were deposited on a low relief, partially vegetated plain and on sabkhas. The layers were separated by an unconformity from the overlying layers composed of near-shore marine to coastal sands. The age of these sands ranged from *c.* 11.5 ka to 8 ka BP. A hiatus separated these transgressive sands from the overlying layers, which were younger than 7.5 ka BP and which were deposited in variable environments, from an inner shelf to a lower deltaic alluvial plain.

The paleo-geographical interpretation of this sequence of deposits by Stanley and Warne (1993) is as follows.

1. The lowest unit of alluvial and sabkha sands (35 ka to 18 ka BP) was deposited on an alluvial plain across which seasonally active braided channels flowed. At that time, the sea had a low level, its coast being located about 50 km further north than at present. Floodplain mud accumulated in ephemeral seasonally dry depressions. Carbonate-rich desert sand and sabkha mud were deposited in the west. The climate during this period was arid.

2. From 15 ka to 8 ka BP, the sea advanced southward, reworking the alluvial deposits. The modern Nile delta began to form *c.* 7.5 ka BP. About 6.5 ka BP, the sea level was some 9 or 10 m below its present level. The river gradient was steeper and the climate was more humid.

3. By *c.* 4 ka BP, the sea level continued to rise, but more slowly. Climate became more arid, flood levels subsided and more distributary channels carried a less-coarse bed load.

4. By *c.* 2 ka BP, sea level had risen to about 2 m below the present level and the delta took on its present configuration. Stanley and Warne explained the fivefold thickening of the Holocene deltaic sequence from west to east by differential subsidence, which resulted in a northeast tilting, connected to a major system of regional faulting.

From the three sets of sections (Foucault and Stanley, 1989; Sneh *et al.*, 1986; Stanley and Warne, 1993), it can be concluded that the lower non-marine sequence of the Last Glacial Period was deposited in an area from which the sea had retreated because of glaciation. At this time, the Nile was at its lowest levels as a consequence of the weak monsoon system. This period was followed, from 15 ka to 8 ka BP, by a rise in sea level caused by global deglaciation. The hiatus between 12 ka and 11 ka, separating nonmarine deposits from the overlying near-shore marine deposits, may represent the Younger Dryas regression. The following hiatus (i.e., before 7.5 ka BP) may be attributed to the mid-Neolithic cold phase. The failure to deposit sediments was probably caused by the retreat of the sea, so marine layers were not deposited, and weak monsoons, so supplies of Nile sediments was also reduced. The modern delta started to form at c. 7.5 ka, which can be correlated with the Upper Neolithic of the Levant, a warm period. Although sea-level rise was relatively slow, the strengthening of the monsoons caused the influx of flood sediments to dominate the reworking of the marine sediments by the rising sea. No Nile deposits evidencing a regressional phase during the Chalcolithic period and EB were observed. Again, this may have been because regression was accompanied by a low Nile, owing to a weak monsoonal system, and, therefore, no deposits reached the lower stretches of the delta. The rise in sea level c. 4 ka BP can be correlated with the transgressional phase of the MB warm period. I believe the rise in sea level c. 2 ka BP, to about 2 m below its present level, represented the sea rise caused by the short warm period from c. 2.3 ka to 2.4 ka BP, which was also observed along the shores of England (Thompson, 1980).

Bell (1971), analyzing Egyptian historical literary sources, found evidence for two drought waves, one between 4.18 ka and 4.15 ka BP and the other between 4 ka and 3.9 ka BP. According to Bell (1971), these severe droughts were responsible for catastrophic famine, especially in upper Egypt. He also claimed that the abandonment of the EB culture in the Mediterranean areas, the collapse of the Acadian empire and the fall of the Samarian Ur III kingdom were all caused by the 4 ka BP climatic crises. Nicholson (1980) mentioned that the Sahara was considerably moister between 6 ka and 5 ka BP than it is now, and a major change towards aridization had appeared by 4 ka BP. Rognon (1987a) found proof of a rapid aridization in north Africa in general between 4.5 ka and 4 ka BP. Butzer (1966) claimed that the beginning of the fifth millenium BP was the last humid period in north Africa.

Stanley *et al.* (2001) have discovered in Egypt's Abu Qir Bay the submerged ruins of two Hellenistic–Byzantine cities, Herakleion and Canopus. The ruins are below 6–7 m of water and are located on the Canopic branch of the Nile delta. These authors believe that the increased supply of sands, which started in the eighth century AD, led to the submergence of the cities. This was the period of the Moslem–Arab warm climate change in the Levant, which brought the strengthening of the monsoonal system over the catchment basin of the Nile, as already discussed.

A recent mean sea-level rise of 8.4 cm was observed at Alexandria from 1944 to 1973 (this was 2.9 mm/year at Port Said).

From 1926 to 1970, there was a mean sea-level rise of 10.1 cm (2.2 mm/year). Part of the rise that has taken place on the Nile delta coast can be explained by the thermal expansion of the upper layer of the oceans, resulting from the observed warming of 0.4 °C in the past 100 years. The other part of the rise may be related to subsidence. It is predicted that the sea level will rise by 37.8 cm and 28.6 cm by 2100 AD over the 1970 level at Alexandria and Port Said, respectively (El-Fishawi and Fanos, 1989).

4.2 THE SAHARA AND THE SAHEL BELT

4.2.1 Contemporary climate

The narrow area influenced by the Mediterranean climate continues westward to the Atlantic Ocean. It widens inland in the mountainous areas because of the altitude effect. In the Atlas region, the marine influence of the Atlantic Ocean is dominant. Along the coastal zone of the Mediterranean, the average annual temperature is 17–18 °C. This is a higher average temperature than that found along the European Mediterranean coastal zone. The annual average amplitude is more than 12 °C, higher than that found in the European coastal area. (In Algiers, the average August temperature is 25 °C and that in January is 12 °C.) Precipitation usually occurs between November and May, with the peak rainfall occurring in the middle of the period. As mentioned above, the precipitation is influenced by the position of the region and by its mountain relief. The amount of precipitation varies from year to year, the annual average falling between 600 and 1000 mm and decreasing from west to east. (The northern Atlas (Tell) mountain chain has annual rainfall of 1000 to 1200 mm.) The characteristic marine-influenced climate is modified as one goes inland from the coast and becomes drier, while the thermal amplitude increases, compared with coastal areas.

The climate of the Sahara desert is an outcome of its geographic location, in an area of active trade winds, and the anticyclonic sub-tropical areas. These two factors have a negative influence on precipitation. In winter, the high-pressure areas come from the Azores anticyclone, which makes contact with the high-pressure area in the central Sahara. During the summer, these two anticyclonic areas move northward. Because of the high temperature of the Sahara, the air masses do not produce much moisture. In most of the Sahara region, the average annual precipitation is less than 20 mm, and it occurs only at irregular intervals. Rainfall may occur in any season, but it is quite possible for no rainfall to occur for several years.

The Khamsin winds are active in the Saharan coastal area. These winds bring large quantities of dust from the desert and deposit that dust as far away as southern Europe. The Saharan climate is characterized by high thermal amplitude, and very reduced humidity.

The average summer temperature is 36–37 °C; the diurnal amplitude is more than 30 °C in the shade and 50 °C in the sun.

At present, nearly all the rains falling in the Sahel zone are monsoonal in origin. In summer (July–August), these rains reach the Hoggar and Tibesti mountains. However, over the central Sahara, the heaviest rains fall chiefly during the intermediate seasons of spring (March–June) and autumn (September–December). Study of the cloud formations over the Sahara also confirms the importance of the intermediate seasons. Interseason rain is linked with the tropical atmospheric depressions, also called Sudano–Saharan depressions or Khamsin depressions, in the eastern Sahara. Rains of this type are often fine and continuous, whereas the monsoon rains are heavy and stormy. At present, such depressions rarely occur in winter, except in the western Sahara but they are also absent at the height of summer, when the ITCZ reaches the Sahara. Schematically, the synoptic situations are as follows.

1. There is an influx of polar air in the middle or upper troposphere above the Sahara, along shallow troughs in the upper westerlies.
2. Frequently, ahead of these cold troughs, undulations occur in the ITCZ, with brief invasions of humid equatorial air. The undulations of the ITCZ could be caused by the action of cold boreal troughs or by monsoonal surges resulting from perturbations in the southern hemisphere. The depressions created by the cold air aloft favor the advection of humid equatorial air. In this season, the movements of these depressions are eastwards or northeastwards. The advection of humid equatorial air is essential for the formation of rain from these atmospheric depressions.

At present, the scarcity of these depressions over the Sahara in winter can be explained by the fact that, during this season, the ITCZ is situated at very low latitudes. However, when there is an interaction between cold troughs and the ITCZ, the trajectory of depressions remains chiefly over the Sudan and Sahel zones.

A study of all rainfall data available since the beginning of the century for Africa north of the equator was carried out by Maley (1977a). Using the method of spatial correlation of annual rainfalls, it can be shown that, in some years, there is a clear opposition between the Sahel and central Saharan zones, either the central Sahara gets heavy rains and the Sahel gets low or average rainfall, or vice versa. The phenomenon is less clear for the eastern Sahara both because data are scarce and because it likely that a trend for the whole Central Sahara would only be apparent over longer periods. For example, only anomalies over 30 years in length have a similar trend through the Mediterranean area. It seems that there are two kinds of wetter periods (optima) for the central Sahara: the first was more frequent in the Quaternary period, with tropical depressions and relatively low temperatures, and the second is characterized by direct monsoon rains and higher temperatures.

4.2.2 Climate during the Upper Pleistocene and Holocene

In north Africa, which still belongs to the westerlies belt, a humid period can be clearly recognized from about 5 ka to 3 ka BP, characterized by a period of geomorphological stability and pedogenesis, producing dark humic soils, quite different from the underlying reddish soils (Rognon, 1987a,b). Alluviation of silts and fine sands occurred during this wet period in many valleys of north Africa. Swamps and little lakes developed in closed basins along the northern Sahara margins (Rognon, 1987a,b).

Pachur and Braun (1980), investigating the paleo-climates of the Libyan desert, central Sahara, found a remarkable difference between west and east in the period from 12 ka to 5 ka BP. There were biologically highly active freshwater lakes in Libya, while in Egypt there were swampy environments with high salt content and active sand transport. The period after 5 ka BP showed a trend to decreasing precipitation. Around 2 ka BP, however, it was more moist although insufficient to cause lake formation.

The climate changes that have affected the northern Sahara during the Holocene were investigated in the Chott-Raharsa basin, Tunisia (Swezey et al., 1999). Dune stabilization and high lake levels characterized the humid periods, while during the dry periods the Chott went through a process of playa desiccation, deflation and dune field construction. The humid periods extended from c. 10 ka to c. 9 ka, from c. 7 ka to c. 6 ka and from c. 5 ka to c. 4 ka BP; between these periods and afterwards the climate was dry. Based on the character of the littoral deposits of the Saharan Atlantic coast laid down during the last 150 ka years, Weisrock (1980) determined climate changes along the littoral part of the Sahara. He found that there was a very dry period (expansion of the western ergs) from 30 ka to 13 ka BP, while from 11 ka BP in the southern part of the Sahara (Mauritania), and from 7.3 ka BP in the north (Morocco), there was a general increase in moisture. This moist period ended somewhat earlier in the south (c. 3 ka BP) than in the north (2 ka BP). The process of becoming arid occurred earlier in the south, became more severe there and continues to the present time. By comparison, in the north, there are still various oscillations, with humans certainly accelerating the process of aridity.

Petit-Maire (1980a) investigated the Holocene biogeographical variations along the northwest African coast (28–19 ° N), where present rainfall is 30 to 50 mm/year. She found evidence for a transgression from 6 ka to 4 ka BP and another at c. 3.5 ka BP. She also found an association of steppe vegetation and animal life with important Neolithic cultural traits, as far north as 21° N. These findings imply that climatic conditions were very different from the current hyper-arid ones, having a minimum annual rainfall of *c.* 150 mm (compared with 28 mm today). By 2 ka BP, the steppe belt shifted back south, perhaps quite rapidly. Petit-Maire (1980b) also reported on Pleistocene lakes in the Shati area, Fezzan (27° 30′ N).

Petit-Maire (1987) summarized her findings in the Tauodenni basin, northern Mali, western Sahara as follows.

1. From 40 ka to 20 ka BP. A humid phase occurred, with Aterian prehistoric artefacts, followed by the Upper Pleistocene arid phase.
2. 10 ka BP. Marshes and lakes formed in the area.
3. From 9 ka to 7 ka BP. An important lacustrine phase was dominant and remains of prehistoric humans and big mammals are abundant. The minimum precipitation was 300 mm. During this period, the Niger flowed into a large interior delta north of Timbuktu. During its major flood, its deposits may have reached an area 300 km north of its present curve.
4. At *c.* 6.5 ka BP. A drier episode was indicated by the presence of layers deposited as a result of evaporation.
5. At *c.* 5.5 ka BP. A new lacustrine episode was recorded. However, the water level was lower than during the Early Holocene. There were no mega fauna and no human settlements north of the 23rd parallel.
6. 4.5 ka. There were sheet flood sediments, evidence of an arid "wadi" regime.
7. By 3.8 ka BP. The lakes had dried up and a sabkha system dominated. People gathered around the water holes and wells in the largest depressions.
8. 3 ka BP. Eolian processes started again to lead to the most severe arid conditions on earth, which continue to the present.

Hoelzmann *et al.* (2000) have estimated the rates of precipitation falling on northwest Sudan in the eastern Sahara during the period of the existence of the lakes, (i.e., from *c.* 9.5 ka to 4 ka BP). They carried out a water balance for a lake, the area of which fluctuated between 1100 and 7000 km^2. In their opinion, the main supply for the lake came from surface flow, while groundwater inflow played a minor role. The rainfall needed to balance evaporation was calculated to be 500 mm per annum for a lake with an area of 1100 km^2 and 900 mm per annum when the lake reached 7000 km^2. At present, the annual mean precipitation is less than 15 mm. The highly depleted ^{18}O values of the lake carbonate reflected, in the authors' opinion, intense tropical summer (monsoonal) rainfall with heavy thunderstorms. Abell and Hoelzmann (2000) concluded that there was considerable seasonality in the annual rainfall, based on isotope retention during the growth of the shells of gastropods in these lakes.

Faure and Faure-Denard (1998) surveyed the climatic changes that the Sahara desert underwent during the Quaternary. During the wet intervals, corresponding to interglacial regimes, inland sand dunes were inactive, flattened and covered with Sudanian and Sahelian vegetation. The main evidence for wet phases comes from laminated diatomites and lacustrine limestone, particularly evident for deposits dated *c.* 30 ka to 20 ka and *c.* 9 ka to 6 ka BP. During the wet episodes, the Sahara was a terrain over which vegetation, fauna and soils were widely expanding, with humans having left their mark everywhere. At least five to ten times more carbon than at present was stored in the total biomass. During the dry phases, eolian conditions of sand and dust movement prevailed, producing areas covered by sands and areas of deflation. During the dry periods, groundwater levels were probably very low. The lacustrine deposits, preserved even in the most arid sectors, testify to Holocene summer monsoon rains regularly reaching north of the Tropic of Cancer at around 8.5 ka BP.

Servant and Servant-Vildary (1980) found that the paleoclimates in the Chad basin seemed to be related to variations in the frequency of incursions of cold polar air to lower latitudes. Changes in the strength of the sub-tropical anticyclonic cells were a major factor controlling late Quaternary climatic changes. In order of decreasing importance, the wettest phases were 11 ka, 9 ka to 8 ka, *c.* 6 ka and 3.5 ka to 3 ka BP. The driest phases were from 20 ka to 13 ka, *c.* 10 ka, *c.* 7.5 ka, 4.5 ka to 4 ka BP and in historical times (i.e., after 3 ka BP). The coldest water temperature, as derived from the diatom flora, occurred from 26 ka to 20 ka BP (dominant temperate diatom flora) and from 12 ka to 7.5 ka BP. Tropical diatom associations, indicating warm water, were abundant from 20 ka to 18 ka BP and from 7 ka BP to the present. From 8 ka to 4 ka BP, Sudan–Guinean and Sudan-type pollen were abundant in what is now the Sahel zone. Indeed, from terrestrial evidence in the Sahel (deflation and incision), it is clear that an arid phase occurred between 8 ka and 7 ka BP, followed by another major humid period lasting from 7 ka to 4 ka BP (Rognon, 1987a; Talbot, 1980). The climate seems to have been less humid, however, than in the period from 10 ka to 8 ka BP. Moreover, unlike the previous wet period, rains were concentrated over a few months of the year and the dry season expanded progressively (Gasse *et al.,* 1980; Rognon, 1987a; Servant and Servant-Vildary, 1980; Street, 1979). Streams were characterized by a braided regime, but perennial flow probably did occur in the major rivers (Talbot, 1980). In the Sahel, following an arid phase recognized in the Chad basin between 11 ka and 10 ka BP (Servant *et al.,* 1976), as well as in other places (Rognon, 1987a), the climate was most humid from 10 ka to 8 ka BP. Fluviatile sediments from this period are characterized by silt and clay accumulations at many sites. The larger rivers of the Sahel were meandering streams with high concentrations of fine, suspended sediments. It is considered that precipitation during this period was not just more abundant but was also more evenly distributed through the year (Rognon, 1976, 1987a; Maley, 1977a; Servant, 1974; Sombroek and Zonneveld, 1971; cf. Talbot, 1980).

Lézine *et al.* (1990) reported on an early Holocene humid phase in the western Sahara according to data from a section from the Chemchane Sabkha (Mauritania, 21° N, 12° W, 256 m above MSL). The age of the lowest part of the section was *c*. 13.5 ka BP. Maximum lake expansion was between *c*. 8.3 ka and 6.5 ka BP, which could be identified by a girdle of stromatolite carbonates. These document a lower-salinity lake, concomitant with the general establishment of Sahel–Sudan vegetation. The Chemchane area exhibits the northernmost (21° N) occurrence of the humid-phase elements that were related to maximum intensity of monsoonal activity during the early Holocene.

Sonntag *et al.* (1980), investigating the environmental isotopes of the fossil groundwater under the Sahara, found that this water was mainly recharged during the long humid period between 50 ka and 20 ka BP, while from 20 ka to 14 ka BP there was a long dry period.

Jäkel (1987) summarized the investigations he and Geyh (Geyh and Jäkel, 1974) had been carrying out on climatic fluctuations in central Sahara during the Late Pleistocene and Holocene. They constructed a histogram of humid and arid climatic phases in the Tibesti mountains and in central Sahara based on changes in the depositional character of the formations, using ^{14}C dating. Layers of a lake, which according to the ^{14}C method dated from 16 ka to 13.5 ka BP, indicate a humid climate during this period. (Although Geyh maintains that these dates are unreliable, as no pluvial deposits are known from the Sahara, I would maintain that the humid conditions were caused by a northern shift of the western African monsoon). Layers in a shallow lake, dated from 12.5 ka indicated a continuously humid period from 12.5 ka to 11.5 ka BP, an arid period from 7.3 ka to 6 ka BP, and a further humid period, from 6.2 ka to 5 ka BP, when a lake reappeared. The period from 5 ka to 4.2 ka BP was very dry. An oscillation towards increased humidity re-occurred between 4.2 ka and 3.7 ka BP. From then onwards, the climate became more and more arid, with an apparently rhythmic pattern of fluctuations. An exception to this pattern occurred between 2 ka and 1.2 ka BP.

These observations and Lézine's data clearly show the negative correlation between the circum-Mediterranean region, influenced by the westerlies, and this area, mainly influenced by the trades, although some extreme shift may have caused anomalies. The influence of the Atlas mountains seems to be similar to that played by the Himalayas with regard to western China.

According to Muzzolini (1986) the west African Sahel, at 5 ka BP, was still under a humid climatic regime, which had begun *c*. 7 ka BP and lasted until 4 ka BP. In south-central Mauritania, the annual rainfall in the period between 5 ka and 4 ka BP appears to have been twice that of today, based on floral and faunal evidence. An arid phase occurred from 4 ka to 3.5 ka BP, during which monsoon precipitation declined sufficiently to allow for local eolian dune formation (Maley, 1977b; Servant, 1973, 1974; Talbot, 1980). This arid interval was followed by the last major humid period in the west African Sahel, which lasted from 3.5 ka

to 2 ka BP (Talbot, 1980). A more arid climatic regime, albeit with fluctuations, has been predominant in the region ever since.

A well-dated core from the Atlantic Ocean, about 200 km off the African coast near Mauritania (core 13289; Koopmann, 1981), revealed a marked and sudden decrease of trade wind speeds *c*. 15 ka BP, as indicated by grain size data from eolian–marine dust deposits entered into the equation of Sarnthein *et al.* (1981). This sudden reduction in speed of the trades above north Africa was remarkably synchronous with the onset of global deglaciation and the decrease of primary plankton productivity in core 12392 (Sarnthein *et al.*, 1987).

Pollen analysis was carried out on Holocene lacustrine deposits sampled every 10–20 or 30 cm in a section of about 7.80 m at Tjeri (13° 44′ N 16° 30′ E) near the center of the great paleo-Chad (Maley, 1977a). The comparison of the Sudano–Guinean element and the Sahelian element curves showed that the periods of climatic optima (relatively wetter phases) were generally out of phase with each other. It also seems that the Sahelian climatic optima have always been synchronous with warming periods during the Holocene, and their deteriorations have coincided with the cool periods. Indeed, the trends of the Sahelian curve at Tjeri – the amplitudes of variations being different – correlated well with the trends that appeared in some curves portraying the evolution of temperature in the northern hemisphere, such as that of Camp Century in Greenland.

Based on environmental responses to climatic change in the west African Sahel over the past 20,000 years, Talbot (1980) deducted the following changes:

1. 20 ka to 12 ka BP: arid, characterized by major dune building;
2. 12 ka to 11 ka BP: humid;
3. 11 ka to 10 ka BP: arid;
4. 10 ka to 8 ka BP: most humid;
5. 8 ka to 7 ka BP: arid;
6. 7 ka to 4 ka BP: humid;
7. 4 ka to 3.5 ka BP: arid;
8. 3.5 ka to 2 ka BP: humid;
9. 2 ka to present: arid.

Some remobilization of older dunes also seems to be occurring under present day conditions. Apart from dunes of anthropogenic origin, which appear almost anywhere in the Sahel, where human interference has greatly reduced the natural vegetation cover, there are active dunes at places along the transition zone into the Sahara.

Baker *et al.* (1995) have performed a synthesis for the paleohydrological conditions during the Late Quaternary for the arid and semi-arid areas of the world based on various studies, some of which have been discussed in this book. For the Sahara and Sahel, they concluded that humid conditions prevailed over these regions and all lakes were at high levels from 40 ka to 20 ka BP. From 25 ka to 16 ka BP, wet conditions persisted over the northern Sahara, while the southern parts dried up. From 16 ka to 12 ka BP, moist conditions prevailed in northern Sahara, while aridity continued in the southern Sahara and Sahel. At 12 ka BP, the lacustrine episode commenced in central and southern Sahara, extending from the line crossing the Tibesti Mountains in the north and covering Mauritania and the Sudan. This phase reached its peak between 9 ka and 6 ka BP.

Maley (2000) explained the lacustrine and fluviatile formations in the Tibesti and other mountains in central Sahara on the basis of an increase in rainfall, connected with the tropical depressions that were formed in the southern part of the westerlies. This movement was linked with activity of the sub-tropical jet stream. These periods with increased rainfall occurred between 20 ka and 12.5 ka BP (with a short pause between 15.5 ka and 15 ka BP).

These and other studies (Roberts, 1989, p. 53; Street-Perrott and Roberts, 1983) support the suggestion that the influence of the westerlies moved southward during cold periods while phases of warming were accompanied by strengthening of the monsoon in central Africa.

A review of palynological and paleo-botanical data from the Sahel for the last 20 ka (Lézine and Casanova, 1989) showed fluctuations in the vegetation zones in tropical north Africa. The migration to the south of the Sahelian pseudo-steppe occurred as far north as 10° N during the last arid episode (18 ka BP). During the Holocene, two abrupt changes in floral landscape occurred, at 9 ka and 2 ka BP. The first was a rapid expansion of humid vegetation zones up to 400–500 km north of their modern position. Latitudes 14–16° N were covered with Guinean and Guinean–Sudanese forest. The second change (2 ka BP) was the disappearance of trees, without any transition, and the beginning of the modern semi-arid environment in the Sahelian zone.

A similar review of pollen and hydrological data for tropical West Africa also showed the two abrupt major vegetation changes mentioned above and, additionally, a front of moist air progressing from south to north during the early Holocene (Lézine and Casanova, 1989). The south Sudanese eco-climatic zone was almost permanently under the influence of humid air masses. The southern margin of the Sahara recorded major hydrologic phases from 9.5 ka to 7 ka BP and from 4 ka to 2.5 ka BP. Moisture conditions from 6.3 ka to 4.5 ka BP were well defined in the Sahelian zone but there was only a mosaic-like pattern in the southern Sahara, reflecting local morphological settings in the three eco-climatic zones. The first recorded evidence of runoff was as early as 12.5 ka BP, whereas modern arid conditions appeared c. 2 ka BP. Regular drizzles, on a year-long basis, occurred from 12.5 ka to 7.5 ka BP. The beginning of strong seasonality in precipitation, and the appearance of a marked dry season, occurred at 7.5 ka BP.

Michel (1980) used sediments from southern Mauritania, on the southwestern Sahara margin to derive climatic changes. At c. 900 BC, it was relatively humid, while drier conditions have prevailed since c. 500 BC, with minor variations: the Middle Ages were relatively humid.

The eastern Sahel region includes the transitional zone between the Sahara and the Sudanese region. The average annual precipitation does not exceed 400–500 mm. The rainy period lasts for 3 to 4 months, during which the effects of the equatorial monsoons are felt. In the hottest month, temperatures vary between 33 and 35 °C; during the coldest month, the temperature is 20–22 °C, (Ashbel, 1938).

The climatic conditions in the monsoon belt of eastern Africa were deduced from ^{18}O and pollen records obtained from an off-shore marine core in the northern Arabian Sea by Van Campo et al. (1982), who concluded that, in general, "Glacial periods were arid in southwestern Asia and at low latitudes, and the northeast trade winds were intensified against a decreased monsoonal flow".

In the tropics, as evidenced by the most detailed Holocene records, incipient interglacial stages were the most humid periods: the southwestern monsoonal flow was intensified over the Atlantic, Africa, the Arabian Sea and India. The insolation peak developed low pressures over the tropical continents and heavy rains resulted. Meanwhile, the middle and high latitudes, more influenced by the northern ice sheets, also had warm, but dry, summers and reached their humidity peak later.

In general, this means that there should be a negative correlation between the southern and the northern Sahara. The former is related from the climatic point of view to the sub-tropical Sahel belt, which, in turn, is mainly influenced by the monsoonal regime. The northern Sahara is related to the circum-Mediterranean area, under the influence of the westerlies.

The Red Sea is also influenced by the climatic regime in the eastern Sahara and this is reflected by the sediments transported to it by its rivers. Thus, the impact of this climate on the hydrology can be derived from investigations of the sediments of this sea. Almogi-Labin et al. (1989) investigated the paleo-environmental events in a Holocene sequence from the central Red Sea, as recorded by pteropoda assemblages. Eight events could be recognized in the biological and sedimentological features of which three were most pronounced.

1. 10 ka to 7.8 ka BP: Water column conditions were unstable. The invasion of normal marine Indian Ocean water caused a sharp decrease in salinity throughout the whole water column, followed by a strong stratification of shallow water depths. Increase in humidity in the surrounding landmass probably caused this stratification.
2. 7.8 ka to 5 ka BP: Similar to present conditions.
3. 5 ka to 2 ka BP: The Red Sea was more oligotreophic than today and the oxygen minimum zone was less developed.

Recent discoveries of fossil-bearing Holocene lake sediments from the eastern Sahara have brought further confirmation and detail to earlier indirect evidence that a major pluvial episode occurred between 9.5 ka and 4.5 ka BP. The botanical records, mainly palynological, show that savanna and desert grassland occupied regions that today are plantless hyper-arid deserts. The indication is that the well-defined latitudinal zonation that characterizes the modern vegetation also existed in the early Holocene, displaced northward by at least 2° (or 450 km). These and other analyses from the Sahara–Sahelian belt imply that the steep gradient of summer precipitation (100 to 440 mm/year), which occurs from 12 to 17° N, was displaced 4–5° northward in the period from 10 ka to 5 ka BP (Ritchie and Haynes, 1987).

Gasse (1980), in his study of late Quaternary changes in lake levels and diatom assemblages on the southeastern margin of the Sahara, reported on a wet period with expanding lakes between 5000 and 2000 BC. There was a dry period of widespread regression from 2500 to 1500 BC and most of the lakes were at low levels, several of them falling below their present level by 1500 BC. From then to the present, the area has suffered a rather dry regime, punctuated by short humid phases (700 BC to 1000 AD: wetter in Ethiopia).

4.2.3 Lakes of eastern Africa

The history of Lake Turkana in northern Kenya was investigated by Owen *et al.* (1982). The lake, which covers an area of *c.* 7500 km^2, is in the semi-arid part of Kenya's rift valley. The drainage basin of the lake is in the Ethiopian highlands. Because it is a closed lake, the fluctuations in its level have an impact on the chemistry of the water. Owen *et al.* (1982) mapped and dated the sediments of the ancient lake outcropping above the shores of the lake. They located three phases of high levels. The earliest reached 80 m above the present level and extended in time from 10 ka to *c.* 7.5 ka BP. There is evidence for a drop in level, to about 10 m above the present level, around 6.6 ka BP, followed by a high level, around 70 m above the present level, that extended to *c.* 4 ka BP. There was a short regression, extending to *c.* 3.8 ka BP, with the level falling some 20 m to about 50 m above the present level, which was followed by a high stand, reaching a maximum high level of 80 m *c.* 3.2 ka BP. Following this stage, the level of the lake started to fall, reaching a minimum *c.* 2.1 ka BP. Immediately afterwards, the level rose a little, to about 15 m above the present level *c.* 1 ka BP, and then fell to its present level.

Gillespie *et al.* (1983) reported on the fluctuations of lake levels in the Ziway Shala in Ethiopia. According to their observations, low levels prevailed before 12 ka BP, from *c.* 10.4 ka to 9.8 ka BP, from 7.8 ka to 7.2 ka BP and from *c.* 4.5 ka to 2.5 ka BP.

Street-Perrott and Perrott (1990) connected these recessions with phases of influx of freshwater into the North Atlantic, from melting ice, which disturbed the monsoonal system.

Comparing these observations with the sequence of paleoclimates of the Levant, it can be seen that the two episodes with low lake levels in the lower Holocene (i.e., from *c.* 10.4 ka to 9.80 ka BP and from 7.8 ka to 7.2 ka BP) in east Africa can be correlated with the two relatively cold periods occurring during the Early and Middle Neolithic in the Levant. This may mean that the cooling of surface ocean water resulted from a global cooling effect rather than the inflow of freshwater from the melting of ice. Street-Perrott and Perrott (1990) connected these recessions with phases of influx of freshwater into the north Atlantic caused by a warm phase, which melted the circumpolar ice sheets. This influx cooled the surface water of the ocean and this, in turn, disturbed the monsoonal system. In my opinion, the mechanism suggested by Street-Perrott and Perrott is true only for the last period, from *c.* 4.5 ka to 2.5 ka BP, while the former periods when the African lakes were at low levels were a consequence of global cooling. Support for a cool global episode during these periods can be found in $\delta^{18}O$ measurements in an ocean core from the Bermuda Rise (EN120 GGC1), which Street-Perrott and Perrott (1990) presented. This curve shows an increase in the $\delta^{18}O$ during these periods, indicating trapping of depleted water by the glaciers. The phase of low lake levels from *c.* 4.5 ka to 2.5 ka BP, which they consider should correspond to the melting of ice, correlates in the Levant with low levels of the Dead Sea and high levels of the Mediterranean Sea, indicating a warm climate. In addition, this conclusion generally corresponds with detection of lighter water in the ocean (i.e., release of light ice water), as can be seen in the oceanic isotope curve. Therefore, it seems that the data from the Levant support the freshwater input explanation, but only for the Middle Holocene episode.

The beginning of a major drought in the Sahel in 1968 has been associated with the global cooling trend, which started *c.* 1964 (Kukla *et al.,* 1977; Newell and Hsiung, 1987). However, the drought also continued after 1977, when global temperatures began to show a warming trend. Zonal mean air temperature for the 300 to 700 mb layer in the tropics clearly rose after 1977 (Newell and Hsiung, 1987).

4.3 TROPICAL AFRICA

4.3.1 Climatic Changes during the Upper Pleistocene and the Holocene

In general, rather arid conditions prevailed over all equatorial regions during the Last Glacial Maximum and landforms were characterized by reduced stream activity. These conditions lasted until *c.* 13 ka BP, afterwards large paleo-floods occurred until 11 ka BP. From this period to *c.* 9.5 ka BP, dry conditions returned. From 9.5 ka BP, the re-establishment of the lowland rainforests started, followed by several wet dry oscillations (Thomas and Thorp, 1995).

4.3.1.a TROPICAL EAST AFRICA

Paleo-climatic records adjacent to India and Africa show that monsoon maxima occurred during interglacial conditions (and coincided with precession maxima and with maxima of northern hemisphere summer radiation). Prell and Kutzbach (1987) used a community climate model to identify the processes causing changes in monsoon circulation. The model indicated that "the regional monsoonal response to glacial age boundary conditions is variable. At 18 ka BP, the monsoon is greatly weakened in southern Asia but precipitation increased in the western Indian Ocean and in equatorial North Africa . . . Both areas show stronger monsoon with increasing solar radiation during interglacial conditions."

Lauer and Frankenberg (1980) found a relationship between plant cover, the number of plant species and annual mean water balance. They concluded that the climate was cold and dry c. 17 ka BP while at 5.5 ka BP it was warm and relatively wet.

Lake levels in Africa and the Arabian peninsula were at their highest from c. 9.5 ka to 5 ka BP, except for an interval around 7.5 ka BP when levels were lower (Street-Perrott et al., 1985). A climatic model (Kutzbach and Street-Perrott, 1985), incorporating changing orbital parameters and surface boundary conditions, simulated a strengthened monsoon circulation and increased precipitation in the northern hemisphere tropics, culminating in the period from 9 ka to 6 ka BP.

Street-Perrott and Roberts (1983) analyzed the data on lake level fluctuations in Africa since the peak of the Last Glacial. They found five major time divisions.

Phase E: Last Glacial to 17 ka BP. Lake levels along the Mediterranean were high, while the lakes inside the continent were low. This situation was explained by a shift towards the south of the belt of the westerlies cyclonic troughs, but it could also reflect low temperatures, which would result in low levels of evaporation. The same is true for southern Africa.

Phase D: 17 ka to 12.5 ka BP. This is a period of low lake levels. Between 14 ka and 12.5 ka BP, an extreme dry period occurred, which was drier even than that of the Glacial Maximum and has affected the whole continent.

Phase C: 12.5 ka to 10 ka BP. At this point, lake levels started to rise. They fell again from 10.8 ka to 10.2 ka BP.

Phase B1: 10 ka to 7.5 ka BP. Almost all intertropical lakes had reached high levels by 9 ka BP. The geographical trend of increasing lake levels was from the equator northward. From 8 ka to 7.5 ka BP, there was a marked regression of many African lakes.

Phase B2: 7.5 ka to 5 ka BP. There was a general rise in lake levels, which reached its peak at 6.8 ka, and lake levels remained high until 5 ka BP.

Phase A: 5 ka BP to present. There has been a rapid decline in lake levels, which continues to the present.

Street-Perrott and Roberts (1983) explained the general trend of high lake levels from 12.5 ka BP onwards, by a general movement of the ITCZ northward. The short but extreme dry periods may be explained by solar events, volcanic eruptions or surges of major ice sheets.

Comparing these fluctuations with the climate changes in the Levant, it is suggested that the last stage of phase C, when there was a general decline in lake levels, would correlate with the cold period of PPN A; the earlier part of phase B1 would correlate with the warm wet PPN B. The short stage of lower lake levels from 8 ka to 7.5 ka BP should be correlated with the Middle Neolithic cold period. The rise in lake levels, starting c. 7.5 ka and reaching its maximum at 6.8 ka BP, correlates well with the Upper Middle and Upper Neolithic warm periods. While no regression of African lake levels occurred during the Levant's Chalcolithic period, the cold and humid period of the EB was clearly reflected in the retreat of the levels of African lakes starting at 5 ka BP. The climatic fluctuations observed in the Levant from then on were not observed in the lakes of Africa. The reason for this may be a lack of observations or a more stable climate since the middle of the Holocene.

4.3.1.b TROPICAL WEST AFRICA

Lézine (1998) examined and interpreted the palynological time series existing for West Africa. The climate of this region is influenced by mobile polar highs, generating anticyclonic cells. These supply the low layer of the fluxes moving towards the meteorological equator, as marine trade winds from north to south and continental trade winds from east to west. While the marine trade winds contribute to the southward advection of cool water in the Canaries current, the continental trade wind is responsible for dust and pollen transport to the sea. A complex structure exists over the meteorological equator. In its lower atmosphere, the dry continental trade winds ride above the humid monsoon, located near the surface, causing westward moving isolated rainstorms. In the middle layers of the atmosphere, the upwelling conditions connected with the ITCZ provide the mechanism that is responsible for the abundant and regular rains. During the year, this zone remains more or less in the vicinity of the geographical equator, while during the summer, the meteorological equator moves northward up to 25° N. As a result, precipitation moves northward but decreases from north to south, causing the zoning of the vegetation according to latitude.

Cores in the Gulf of Guinea and off the western coast of Africa showed two well-defined pollen maxima c. 15 ka and c. 10.3 ka (Younger Dryas), and a minor peak at 7 ka BP. The first two peaks were characterized by Saharan type pollen and high amounts of dust-derived deposits. In one of the cores, there was an increase in the percentage of pollen of steppe origin, and it also contained assemblages of foraminifers, indicating lower sea temperatures. The pollen in the layers dated from the Holocene came from the nearest tropical forests. Consequently, it can be concluded that, during the cold periods of the Last Interglacial, the continent went through a dry period, dominated by northeasterly trade winds. During the Holocene, these severe conditions did not recur except *c*. 7 ka BP.

The scarcity of data makes it difficult to reconstruct the rainfall patterns determined by climatic variations during the Holocene. It seems, however, that the 400 mm isohyet was placed at about 20° N during the early Holocene. Afterwards, from 6 ka until 4.5 ka BP, it decreased to 300 mm at Oyo Suan.

Southward, in the humid Sahel ecosystems, an abundance of lakes was characteristic of the environment between 9 ka and 8 ka BP. At *c*. 7.5 ka BP, a well-marked seasonal regime of precipitation is evidenced by the pollen assemblage. Increased rates of precipitation and high lake levels were manifested between 4.5 ka and 2 ka BP. Afterwards, conditions in the Sahel changed to the present situation. The equatorial rain forest did not go through any marked changes during the Holocene, except for a short period of dryness *c*. 3 ka BP.

Preliminary pollen data from sediments from Lake Bosumtwi in southern Ghana (west Africa), dated between *c*. 15 ka and 8.5 ka BP, showed the presence of *Olea hochstetteri* pollen, at 3–8%. This demonstrates the spread of this typical mountain species to low altitudes. Heavy cloud cover lasting most of the year (low clouds of stratiform type and fog) may have caused this spread. It may have coincided with a lowering of the mean temperature by at least 2–3 °C (Maley and Livingstone, 1983).

From a pollen sequence recently obtained in the mountains of the Zaire/Nile divide in Burundi, one gets a picture of climatic changes during the Upper Quaternary for the region south of the equator (Bonnefille *et al.*, 1990). The record indicates an average temperature decrease of 4 ± 2 °C for the glacial period between 30 ka and 13 ka BP. Moreover, the curve seems to be consistent with the evolution of past vegetation, indicated by the same fossil pollen record. Before 30 ka BP, the cooling was consistent with the occurrence of mountain conifer forest on the Burundi highlands (>2200 m), which are now tropical rain forest. There, the lowest recorded temperature occurred between 25 ka and 15 ka BP (Bonnefille and Riollet, 1988). Prior to 30 ka BP, climatic conditions were colder and drier than now. Between 30 ka and 15 ka BP, dry and cool conditions prevailed. More humid conditions started after *c*. 13 ka BP. Bonnefille *et al.* (1990) correlated short-term humid episodes *c*. 21.5 ka BP with high lake levels in eastern Africa and with sapropel layers in the Mediterranean Sea and Atlantic Ocean. Examining the section for the Holocene, although ^{14}C datings are scarce, it can be seen that humid conditions during this period, resembling those of 21.5 ka BP, started *c*. 6.7 ka BP and reached their peak *c*. 5 ka BP.

Finney and Johnson (1991) investigated the paleo-limnology of Lake Malawi, south of the equator. They found that low lake levels persisted from 10 ka to 6 ka BP, which they attributed to lower levels of precipitation. The situation changed after 6 ka BP, when the lake level rose as a result of an increase in precipitation. Upwelling and the lowering of the chemocline may be evidence of lower temperatures and stronger winds at 3.5 ka BP. At *c*. 2 ka BP, conditions became more variable. These results do not agree with observations in tropical Africa north of the equator, which indicated higher water tables, that is, moister conditions (Butzer *et al.*, 1972; Rognon 1987a; Street and Grove, 1976). As phenomena similar to Lake Malawi were reported for southern lakes that were closer to the equator, Finney and Johnson (1991) suggested that a sharp climatic boundary, located near 9° S, divided the lakes to the south and north of this line.

4.4 SOUTH AFRICA

4.4.1 Contemporary climate

The present climate of southern Africa is determined by its position within the sub-tropical anticyclonic belt. This belt is divided into two subsystems or cells, i.e., the south Atlantic and Indian Ocean cells. To the south lie the circumpolar westerlies and their associated cyclonic low-pressure systems. The north-to-south movement of the ITCZ influences the climate north of the 20° S line.

Another important factor deciding the climate, especially the distribution of rains over southern Africa, is the regime of currents in the two oceans flanking the region. The cold Benguela current, which flanks the west coast, flows northwards while the warm Agilhas current, which flanks the eastern coast, flows towards the south.

These two currents, as well as the topography of the country, decide the quantity of precipitation, which is also affected by the movement of the anticyclones and cyclones. In summer, the Indian Ocean sub-tropical moves southeastward, while a weak thermal low-pressure cell develops over the central interior of the country. This enables the formation of convectional cells (local as well as synoptic) over the central and eastern part of the country, causing summer rains. The amount of rain is up to 1000 mm in the east, decreasing gradually towards the west. Along the western coast of southern Africa, mean annual precipitation is below 100 mm.

During the winter, the regime is anticyclonic. High pressure over most of the country deprives it of the humidity of the Indian Ocean. However, a westerlies cyclonic system over the southwestern part of the country brings humidity and winter rain to this part of Cape Province.

Wet and dry spells in the summer rainfall zone are influenced by the regime of pressure over the region. Wet spells are associated with the development of lower pressure over southern Africa and increase in pressure over the south Atlantic. Dry periods are associated with the expansion of the westerlies northward and the weakening of the easterlies. During these dry spells, pressure rises over South Africa and advection of moisture decreases. The increase in temperature gradient between the tropics and the Antarctic causes the cyclonic storm tracks connected with the westerlies to move northward, leading to an increase in winter rains in the southwestern Cape Province.

4.4.2 Climate conditions during the Upper Pleistocene and Holocene

Van Zinderen Bakker (1976) proposed that, during the Last Glacial Period, cool and dry conditions prevailed in north and east South Africa, while cold temperatures, strong winds, and wet winters prevailed in the southern section.

The same author maintains that, during the interglacial periods, pluvial conditions prevailed in the eastern part of South Africa in summer, while dry westerlies prevailed along the western and southwestern coasts. During winter, cyclonic storms penetrated the southwestern regions, while the high-pressure conditions caused dryness in the eastern part of South Africa.

Nicholson and Flohn (1980) suggested that, during the Last Glacial Maximum the ITCZ migrated southward, causing most of southern Africa, except the southern tip of the 30° S line, to be drier than at present. During the early part of the Holocene (10 ka to 8 ka BP), the area north of the 25° S line was more humid than today, while the areas south of it were drier.

Heine and Geyh (1983) suggested that, during the Last Glacial Period, strengthening of the circulation pattern caused stronger trade winds to bring winter rains as far as the southern Kalahari, and summer rainfall over the Kalahari, but not as far as Namibia. During the Post Glacial Period (17 ka to 15 ka BP), eastern circulation caused most of South Africa to be semi-humid, and to benefit from summer as well as winter rains. At c. 12 ka BP, the circulation weakened and semi-arid conditions prevailed in eastern South Africa, where rains fell only during summer. Western South Africa was arid. Some climatologists have pointed out the impact of the ENSO, in which the southern Indian Ocean high-pressure cell is displaced northeastward. This would have caused much drier conditions in the summer rainfall regions of southern Africa, and wetter conditions in the regions of winter rain. Such conditions may have prevailed during the Last Glacial Maximum.

Johnson et al. (1997), on the basis of ratios of carbon, nitrogen and oxygen isotopes from ostrich eggshell at Equus Cave, derived distribution of types of vegetation (C3 and C4) and thus rainfall and temperatures for the last 17 ka. They concluded that paleo-temperatures were at a minimum between 17 ka and 14 ka BP and reached their maximum in the Late Holocene. At 17 ka BP, mean annual precipitation was at a minimum. It increased steadily to modern values by c. 6 ka BP and remained relatively unchanged until the present.

Tyson (1986) suggested that cool periods are generally correlated with enhanced climatic instability, leading to greater extremes of climate. There appears to be general agreement that, during the colder periods of the Upper Pleistocene and Lower to Middle Holocene, the climate was wetter and windier in the winter rain regions, which extended further north and east of the present boundaries, while the summer rain regions were cooler and drier. Periods of warmer climates may have caused higher rainfall in the summer rain regions, even promoting some rain in the winter rain regions, which may have become warmer and drier.

However, the construction of a general paleo-climate curve for the sub-continent of South Africa is difficult to carry out because of contradictions in the conclusions drawn by different research groups. I assume that these contradictions arise through the many instances of subjective interpretation by the various investigators, especially where reconstruction was based on the evidence of paleo-ecological assemblages, such as bones or pollen. These assemblages may have been influenced by local factors as well as climate.

Partridge et al. (1990) reconstructed the climatic fluctuations for the different regions of southern Africa on the basis of paleo-environmental proxy-data. They divided the sub-continent into five paleo-climatic regions.

1. The southern and western Cape, coinciding with mainly "Mediterranean" or winter rain climate;
2. The Karroo semi-arid, mostly summer rain climate (80%);
3. The Kalahari arid summer rain region;
4. The Namib hyper-arid zone;
5. The high precipitation region of eastern South Africa.

Correlating the data from these different regions shows that during the Last Glacial Maximum, 21 ka to 17 ka BP, all the regions, except the Kalahari (which was humid), were characterized by low temperatures (about 5–6 °C lower than at present). Warming proceeded gradually from 17 ka to 12 ka BP, causing wetter conditions in all regions. From 12 ka to 10 ka BP, the warming trend continued. In the eastern and southeastern parts of the sub-continent, the humidity was at the same level as at present, but the Kalahari and Namib deserts became dryer.

In the southern Cape Province, a region characterized by rains during all the seasons, a good sequence was derived for the upper part of the Holocene from a speleotheme taken from Cango Cave (Talma and Vogel, 1992). The paleo-temperatures were calculated from the ^{18}O content of the carbonate. In order to do this, one must know the $\delta^{18}O$ of the water in the past. This was obtained by comparing the $\delta^{18}O$ content with that of the Uitenhage artesian aquifer, located about 350 km east of Cango Cave. The

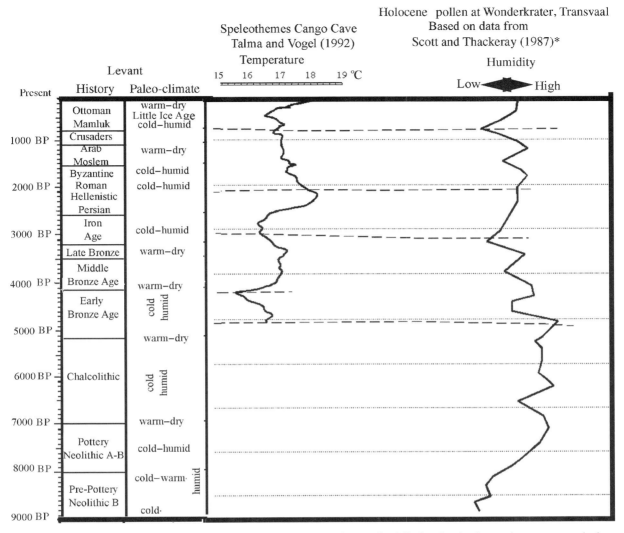

Fig. 4.2. Climates during the Holocene in southern Africa. *Adjusted to scale and streamlined (3–5) points by the running average method.

information derived from the speleotheme regarding temperatures in the Uppermost Pleistocene areas conformed with what would have been expected. There was a temperature decrease from *c*. 30 ka BP, reaching a minimum value between 19 ka and 17 ka BP (reaching about 10 °C compared with 18 °C for the most recent sample).

Afterwards, up to 13.8 ka BP, the temperature increased (at about 14 ka BP, it reached a maximum of about 14 °C). At this time, precipitation of carbonates stopped and was not renewed until 5 ka BP. The range of temperatures from this period to the present is presented in Fig. 4.2.

The ^{13}C contents of the stalagmites formed during the Holocene was higher than that during the Pleistocene. While the latter varied between $-9\%o$ and $-11\%o$ (where minus indicates lower than a standard value), that of the Holocene (since 5 ka BP) started at $-9\%o$ and increased quickly to reach a peak of $-4\%o$ at *c*. 2 ka BP and $-6.4\%o$ at present. The low ^{13}C content of the stalagmites during the Last Glacial Period is a function of the dominance of C3 type vegetation (18%) owing to lower temperatures. Today, the vegetation is composed of more than 60% C4. The slow increase of ^{13}C from 5 ka BP ($-9\%o$ to $-4.9\%o$) does not correspond with the temperature fluctuations curve. Talma and Vogel (1992) suggested that this difference between the ^{13}C and temperature curves could be explained by the fact that the rainfall at 5 ka BP was more of a winter rainfall type than that of today, which is an all-season rainfall pattern.

Issar (1997) suggested that the absence of deposition of calcium carbonate from *c*. 13 ka to 5 ka BP was caused by winter rains starting to decrease and summer rains to increase as temperatures started rising. The vegetation, being dominantly C3, absorbed and transpired all the water infiltrating the subsurface. Only after the ecosystem changed into a floral assemblage, dominated by C4 plants, which do not have deep roots, did recharge to groundwater, and thus drip water, restart.

A similar cessation of stalagmite formation can be found in Boompleas Cave (Deacon *et al.,* 1984; Deacon and Lancaster, 1988). This cave, which is in Cape Province, is an enlarged opening of a fissure that formed a large domed rock shelter. The sequence is rather continuous down to *c.* 80 ka BP. The Last Glacial Maximum, dated between 22 ka and 17.8 ka BP, was characterized by more angular debris in the deposits, presumably as a result of front type erosion, and a pollen assemblage characterized by a low diversity of species. By comparison, the Holocene samples have a high diversity of species. The major change in vegetation from the Glacial Period to inter-glacial in this region was in species diversity, with a substantial increase occurring after *c.* 14 ka BP.

A palynological and paleo-botanical study of the site (Scholtz, unpublished data, cited by Deacon and Lancaster, 1988) concluded that, during the Last Glacial Maximum, the climate was cold and dry throughout the year. The climate was very wet from *c.* 14 ka to 12 ka BP. Total annual precipitation was high and was distributed throughout the year. From 10 ka to 9 ka BP, the climate became a little drier, somewhat cooler and more seasonal than the preceding period. At *c.* 6.4 ka BP, there were strong xeric conditions, probably with little or no rain in the warmer months, and with long, dry, hot summers. From 2 ka to 1 ka BP, there was a decrease in rainfall, especially during the warmer months (October to February). This resulted in a contraction of forest vegetation. Lower temperatures may also have been experienced.

A stalagmite formed in Boompleas Cave, starting during the Upper Pleistocene (later than 80 ka and before 40 ka BP). It ceased to form *c.* 14.2 ka BP and, unlike that in Cango Cave, did not rejuvenate. The higher percentage of trees during the shift from the Uppermost Pleistocene to the Lower Holocene could explain the absence of groundwater recharge of the sub-surface and thus into the cave (as in Cango Cave).

Another source of information on the climate in Cape Province during the Uppermost Pleistocene and Lower Holocene is to be found in the noble gas composition of the water in the Uitenhage artesian aquifer, northeast of Port Elizabeth and *c.* 350 km east of Cango Cave. As gas solubility is a function of the temperature of the water, Heaton *et al.* (1986) measured the concentration of nitrogen (N_2) and argon in the water. The samples, dating between 28 ka and 15 ka BP, indicated mean temperatures of about $14 \pm 1\,°C$. This means that they were 5 °C lower than contemporary mean temperatures in this region. The mean temperature for the period from 9 ka BP to the present was calculated as 19.5 °C.

Another paleo-climatic Holocene sequence was derived from a core into a peat deposit formed by a spring in Wonderkrater in the Transvaal, which is in the summer rain eco-zone (Scott and

Thackeray, 1987; Fig. 4.2). At present, the area receives between 400 and 600 mm of precipitation per annum. The palynological sequence from bottom to top was:

1. Earlier than *c.* 34 ka BP: woodland with expanded montane forests;
2. After *c.* 34.4 ka BP: Kalahari-type savannah with restricted montane forest;
3. At *c.* 25 ka BP: open grassland with more podocarpus forest;
4. At *c.* 25 ka to 11 ka BP: mainly open grassland and restricted montane forest;
5. At *c.* 11 ka to 9.5 ka BP: open grassland with much reduced virtually absent montane forest;
6. 9.5 ka to 6 ka BP: Kalahari-type savannah;
7. 6 ka to 4 ka BP: savannah with broad-leaf element;
8. 4 ka to 2 ka BP: upland bush land type with restricted montane forest;
9. 2 ka to 1 ka BP: more or less the same;
10. 1 ka BP to present: bush land with restricted montane forests.

A multivariate analysis of the pollen assemblages produced two curves, one interpreting the pollen data for moisture changes and the other for temperatures. Comparing these data curves with that of Cango Cave (Talma and Vogel, 1992), one can say that there is a rather good correlation between dry periods on the Wonderkrater curve and cold periods on the Cango Cave curve, while there is a greater variance between the moisture curves from the two sites.

As the evidence from Cango Cave is more direct, it is suggested that only the dry–humidity curve from the Wonderkrater data should be adopted. We may conclude that during most of the Holocene – on the basis of the Cango Cave data – the warm periods were more humid in the Transvaal, while cold periods were dryer.

A more general conclusion can be drawn regarding the regions in summer rains in South Africa: during warm periods these regions will enjoy higher rates of precipitation and vice versa.

This conclusion is supported by an analysis of the stable isotope variations and layer structure of a section spanning 3000 years from a stalagmite at Cold Air Cave, in the Northern Province of the Republic of South Africa (Holmgren *et al.*, 1999). The darker colored layers of the stalagmite, a product of mobilization of organic matter from the soil, were also characterized by a higher ratio of heavy ^{13}C and ^{18}O isotopes. This results from a warm humid climate of summer rains. Such layers characterized the period from 900 to 1300 AD. Lighter coloring and depleted isotopic composition, denoting cool and drier climates, were characteristic for the period from 1300 to *c.* 1800 AD, the Little Ice Age.

Support for this conclusion can also be found in the profile of the Pretoria saltpan, which is the infilling of a meteor impact crater (Partridge *et al.,* 1993). The analysis of the pollen assemblages showed that warm and moderately cool conditions were wet, while a cool climate equated to dry conditions. The resolution for the Holocene was not detailed enough and no pollens were found between *c*. 30 ka and 8 ka BP, most probably because of bad preservation conditions. The few samples that were analyzed showed that rather cool conditions prevailed *c*. 7.2 ka BP; from 4.6 ka to 4.4 ka BP, the climate was warm, and *c*. 2.3 ka BP, it was cool. The climate later warmed up, reaching its current maximum.

Lancaster (1979) found evidence for a widespread Late Pleistocene humid period in the Kalahari desert from 21 ka to 14 ka BP and a sub-humid period from *c*. 9.7 ka to 6.5 ka BP.

Baker *et al.* (1995) concluded that the period from 25 ka to 16 ka BP was a period of desiccation of the deserts of southern Africa. In the Kalahari, a humid period started *c*. 17 ka BP and persisted also during the early Holocene; the Namib desert became dry during the Holocene.

Tyson and Lindesay (1992) gave the pattern of climate changes for 0–1810 AD in South Africa:

1. 100–200: cool;
2. 200–600: warm;
3. 600–900: cool;
4. 900–1300: warm;
5. 1300–1500: cool;
6. 1500–1675: warm;
7. 1675–1780: cool;
8. 1790–1810: warm.

Based on archaeological evidence, Huffman (1996) found that the warm period, which was also humid, extended from 200 to 600 AD (local Early Iron Age). It led to the extension of settlements in the Central Transvaal and in Botswana, reaching the edge of the Kalahari desert. Another phase of occupation of the fringes of the Kalahari was *c*. 850 to 1350 AD (local Middle Iron Age). The Little Ice Age, which started *c*. 1300 and lasted until 1780 AD, with a warmer interval between 1500 and 1675 AD, caused the desertion of the agricultural settlements in the west and shifted the settlement to Zimbabwe's southeast escarpment, which had relatively more rains than the western region because of its warm and wet climate.

A record of the changes of the vegetation of the Kalahari was obtained from investigating a core drilled into a speleotheme in Drotsky's Cave, Botswana. The core, which contained pollen, was dated by the uranium-series method (Burney *et al.*, 1994). It was found that from *c*. 10 ka to 7 ka BP the site was surrounded by arid grassland with trees, which were adapted to dryness. From *c*. 7 ka to 6 ka BP, the assemblage of pollen indicated wetter conditions,

which prevailed until 3 ka BP although a dryer interlude occurred between 4 ka and 5 ka BP. In general, it can be concluded that the changes in Kalahari vegetation during the Holocene were slight relative to other regions in Africa.

As discussed above, the aridity of southwestern Africa is connected with the upwelling of cold waters of the Benguela system. In its southern part, this system is dominated by a strongly annual regime. In the more extreme south, the sea surface temperatures are also influenced by episodic warm Agulhas current intrusions from the Indian Ocean. Cohen *et al.* (1992) reconstructed the Holocene history of this upwelling through an analysis of ^{18}O content and the calcite to aragonite ratio in the shells of *Patella*, found in shell middens in archaeological sites on the coast of southwestern Africa. The calcite to aragonite ratio increases with an increase in temperature. They found three discrete episodes of ^{18}O enrichment in the shells, corresponding with lower aragonite ratios: evidence of glaciation episodes as well as of colder water. The times of these episodes were between 11 ka and 10 ka, between 4 ka and 2 ka and between 0.75 ka and 0.4 ka BP. These authors correlate the first episode with the Younger Dryas cold period, the second with a period of glacier expansion in the northern hemisphere, observed by Lamb (1982), and the third episode with the Little Ice Age. If the diagram by Cohen *et al.* (1992) is compared with the Levant data, it can be said that the first episode can be correlated with the cold period of the Early Neolithic, the second with that of the Late Bronze and Early Iron Age, while the last one can be correlated with the Crusader and Little Ice Age cold periods.

A group of scientists from the Climatic Research Unit of the University of East Anglia at Norwich, UK has carried out a comprehensive investigation on the potential impacts of the expected climatic warming on southern African environments, natural as well as human (Hulme, 1996). The study included an investigation of the impact of past climates as well as of recent trends of change. These data were used in a simulation using a composite computered climate model, which linked results from a general circulation model with a simple climatic model. The running of the model enabled a forecast to be made of a future "core" scenario for South African climate in the 2050s.

In general the "core" scenario forecast is in accord with my conclusions and it can serve as a conclusion for this chapter. This scenario sees modest drying over large parts of the southern part of Africa (south of latitude 15° S) of *c*. 5%, except for the southwestern Cape Province, Zimbabwe and the Transvaal, where the drying may even reach 10%.

5 Climate Changes over western USA and Mexico during the Holocene

5.1 SOUTHWESTERN USA

5.1.1 Contemporary climate

The precipitation regime of the southwestern USA (Fig. 5.1: south of latitude 40° N) is a function of the interplay between the westerlies system over the Pacific and the monsoonal system over the Gulf of Mexico. It has two rainy periods. The summer rains result mainly from monsoonal air masses, which originate in the Gulf of Mexico, and air masses coming from the Pacific. In addition, the height of the Colorado plateau magnifies the thermo-synoptic contrast between continent and sea, forcing air masses to rise towards the low-pressure area over the plateau.

California has a moderate climate with an average annual precipitation in excess of 1000 mm. During the summer, temperatures vary between 27 and 28 °C; in winter, between 7 and 8 °C.

On the Pacific coastal plain (in Mexico), the climate is dry and very hot. Here, cold winds from the northwest blow for about 8 months of the year. During the summer, southwesterly winds bring torrential rains.

The factors that play a role in deciding the relative influence of the two systems (i.e., the westerlies and the monsoons) are the circumpolar vortex (the strength) of the sub-tropical westerlies, and large-scale anomalies in temperatures of the sea surface, primarily those associated with the ENSO and the NAO (Hughes and Graumlich, 1996). The climate of California is especially influenced by the California current. This current is a branch of the southward flowing system of currents of the northeastern Pacific. The southward flowing waters are cold, which is a factor in the aridity of this geographical region. However, from the fall to spring, which is the rainy season, strong northerly winds, which strengthen the current, cause strong upwelling, which pulls in warm water from the south.

5.1.2 Climate changes during the Upper Pleistocene and Holocene

Davis (1995) has built a database of more than 1000 samples from over 50 sources. By applying the analogue technique, he reconstructed the paleo-climates of three sites: the Montezuma Well in Arizona, Exchequer Meadows in California and Rattlesnake Cave in Idaho.

Examining the Montezuma Well paleo-temperature curve, one can divide the Holocene sequence into three stages.

1. The early cold stage, 10 ka to 8 ka BP, in which the range of temperatures was below 12 °C level (c. 10 °C).
2. The mid-warmer stage, 8 ka to c. 2 ka BP, when the range of temperatures was above 12 °C with a few peaks approaching 15 °C.
3. The upper colder stage, from 2 ka BP to present, with temperature fluctuations around 12 °C, with four minimum levels, of which two (at 1.6 ka BP and one in the twentieth century) were just below the 10 °C level.

Comparing the temperature curve with the precipitation curve at the same site, it can be seen that during stage 1, precipitation was rather high, exceeding 400 mm, and reached extremes of more than 600 and even 700 mm during short periods. By comparison, stage 2 was a warmer period and precipitation ranged between 300 and 400 mm, except for a short period around 7 ka BP, when it approached the 500 mm level. During the upper stage 3, the range was more or less similar, although it has had a trend of increasing precipitation. From this, one would conclude lower temperatures corresponded with higher precipitation levels in the Holocene in Arizona.

Evidence from Exchequer Meadow in California indicated that changes in temperatures during the Holocene were less pronounced here than in Arizona. Temperatures did not go below 8 °C and did not exceed 9 °C during the whole of the Holocene, except during stage 1, which was a little cooler. During Stage 2, temperatures reached 9 °C, while Stage 3 was somewhat cooler. Precipitation at the beginning of this stage was rather low, c. 600 mm, but has increased more or less constantly up to the present, reaching an average of about 1200 mm in recent decades. From these data, it is not possible to show a direct relationship between temperature and climate in California for the whole Holocene.

Fig. 5.1. Map of southwestern USA.

The pollen curve from Rattlesnake Cave, Idaho showed a rather warm phase during stage 1, declining towards 5 ka BP and then declining again *c*. 2 ka BP (i.e., during stage 3). Levels of precipitation showed a marked correspondence of low temperatures and higher precipitation, and vice versa.

From the palynological time series Davis (1995) could discern three stages of climate changes for the Holocene in the western USA. In the southern region, both temperatures and temperature impact on precipitation were very variable, while there was greater consistency in the north. This may be because the climate in the south was influenced by more factors than that in the north.

Davis (1992) investigated the pollen assemblages deposited in the San Joaquin marsh, situated near the sea in southern California. The pollen record traces the changes in the sea levels, as the marsh became salty during high sea levels and halophytic taxa dominated. When the sea retreated, freshwater-type vegetation dominated. Davis (1992) correlated the sea retreat episodes with periods of global cooling. These occurred, according to this reconstruction, at 3.8 ka, 2.8 ka, 2.3 ka and after 0.56 ka BP.

Hughes and Graumlich (1996) used the extensive dendroclimatic data existing for the western USA to reconstruct the precipitation for the Great Basin (Fig. 5.1) from *c*. 8 ka BP (Fig. 5.2). This reconstruction showed that the annual precipitation was rather uniform, ranging around 200 mm. Despite the generally uniform nature of the precipitation record, droughts and periods of above average precipitation lasting several centuries still occurred. Prolonged droughts were more persistent during the middle of the first millennium BC, as well as in the sixth and eighth centuries AD. Generally speaking, the conclusions of Hughes and

Graumlich (1996) are in general agreement with the conclusions of Davis (1992), with regard to the precipitation for the Montezuma Well area in Arizona. However, according to Davis, the precipitation from 8 ka to 6 ka BP was somewhat higher.

Graumlich and Lloyd (1996) also reconstructed the paleoclimatic fluctuations since *c*. 3.5 ka BP for the Sierra Nevada based on dendrochronological data. They observed periods of extended droughts during 800–59, 1020–70, 1197–1217, 1249–1365, 1443–79, 1566–1602, 1764–94, 1806–61 and 1910–34 AD. On the basis of records of sub-alpine tree growth and density, they concluded that between 1450 and 1850 AD temperatures were below the long-term average. Yet, as a general conclusion, they claimed that "the dendroclimatic records, if interpreted without reference to other proxy data sources, imply that climate in the late Holocene has been stationary".

Leavitt (1994) observed a strong depletion in $\delta^{13}C$ in Bristlecone Pine tree rings in the White Mountains, from 1080 to 1129 AD, which would suggest a period of abundant soil moisture allowing the stomata of the tree to remain open. In general terms, the series of drought years identified by Graumlich and Lloyd (1996) on the basis of dendrochronological data is in good agreement with Leavitt's findings based on $\delta^{13}C$ data.

Lloyd and Graumlich (1997) found that the tree line in the Sierra Nevada was at a low altitude from *c*. 1100 to 600 BC; it was low for a relatively short period *c*. 100 BC, declined sharply *c*. 1100 AD and remained at that level from *c*. 1500 to 1600 AD. One can assume that these declines were a result of cooler periods and vice versa.

Comparing the precipitation curve of Graumlich and Lloyd (1996) with the tree-line (temperature) curve of Lloyd and Graumlich (1997), and taking into account the fact that their

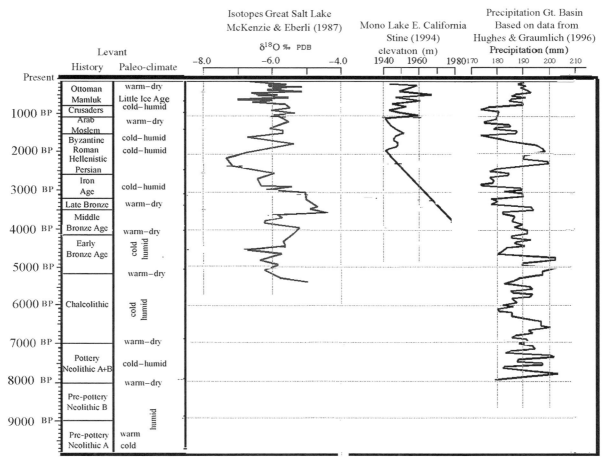

Fig. 5.2. Paleo-hydrology of southwestern USA.

dating is rather accurate, one can conclude that changes in temperature correlate differently with changes in precipitation in different periods: during certain periods with a high-altitude tree line (i.e., periods of warmer temperatures), precipitations were higher, while during other periods, the opposite situation occurred (when there were lower tree lines, indicating lower temperatures, there was also higher precipitation), especially during the last millennium.

Comparing Hughes and Graumlich's (1996) reconstructed precipitation curve with the levels of Lake Mono (Stine, 1994; Fig. 5.2), a rather good correlation can be found between periods of low precipitation and ones of low lake levels (assuming Stine's dates have a 100 year inaccuracy range). This is also consistent with the conclusions arrived at by Graumlich (1993).

Somewhat similar results are to be obtained from the work of Fritts and Shao (1992). They reconstructed the climate for the western USA starting in 1600 AD, based on calibration with data from the period for which instrumental data are available, namely from 1918 to 1961. Altogether, the reconstruction showed low-frequency variations, including the period of the Little Ice Age. Low temperatures and higher precipitation in the northern

parts of the western USA and warmer and dryer conditions to the south characterized the period from 1602 to 1636. Around 1667, the reconstruction of climate showed a uniform decline over the western USA, except for the California–Dakota area, where precipitation seems to have increased because of strengthening of rainstorms over the California–Dakota storm track. Overall, the reconstructed paleo-climates for the last 400 years show considerable variability in all factors concerning differences in the trends of warming and cooling, as well as in temperature impact on precipitation, in the different parts of the western USA. This indicates that during this period, which was characterized by warmer temperatures over most of the western USA (except the Great Basin), there was an increase in precipitation in the southwest, diminishing in the more northern part of the western USA. This trend was connected to a period of low pressures at sea level over the entire North Pacific and especially over the North American Arctic.

The reconstruction of the climate of southwest USA by Davis (1996), was based on percentages of aquatic types of pollen. He emphasized two warm periods: the Medieval Warm Period, c. 1000–1200 AD, and the early Holocene, c. 8 ka to 10 ka BP. During these periods, the monsoonal summer precipitation in the southern part of the American southwest was higher, while in the northern area, it was lower. The more humid climate had a positive impact on the socio-economy of the agricultural native American societies of Arizona that started to grow c. 500 AD, reaching a climax 1100–1200 AD.

Petersen (1994) found a similar positive influence of a warm climate from c. 900 to 1300 AD on the socio-economic system of the Anasazi-Pueblo Native American people in the southern Rocky Mountains. He based his conclusions on the correlation between the archaeological data portraying the expansion of the farming societies and proxy-data, related to the movement of the lower spruce forest border in one site and the timber line at a second site. At the first site, Beef Pastures, the data were obtained by comparing ratios of spruce pollen to pine pollen, while at the second site, Twin Lakes, the data were based on the ratios of conifer pollen to non-arboreal pollen.

Davis (1994) also showed the increase in summer precipitation, caused by strengthening of monsoonal rains during warm periods, by comparing palynological indicators of lake levels with Petersen's (1994) conclusions. He suggested a correlation between these events and increased solar activity.

Scuderi (1990), who investigated tree ring variations at Cirque Peak in the southern part of the Sierra Nevada in California, found a marked correspondence, on the decadal level, between tree ring variations from the temperature-sensitive upper timberline sites in the Sierra Nevada and the sulfur-rich aerosols recorded in the Greenland ice cores. The rate of decrease of temperature was in the order of 1 °C for up to 2 years. He suggested that clusters of volcanic events may have served as triggers of climate changes but there may have been significant spatial variations in the actual changes occasioned by these eruptions. While some regions may have become cooler, others became warmer. He showed that periods when extremely narrow tree rings were produced in the temperature-sensitive sub-Alpine tree *Pinus balfuriana* were related to major volcanic eruptions, recorded historically or in ice cores. These periods also coincided with periods of the advance of glaciers in the Sierra Nevada. The most pronounced periods were 500 BC to 50 AD, 150–250, 540–640, 800–950 and 1600–1890 AD. Intervals of decreased volcanic activity coincided with few or no dated glacial moraines and a minimal number of wider rings; these periods occurred in 950–500 BC and 50–150, 300–450, 600–750, 1000–1250 (Medieval climatic optimum) and 1890–1980 AD. Scuderi (1993) reconstructed the mean June to January temperature at the timberline site for the last 2000 years based on the tree rings. He observed a 125 year periodicity, which may be linked with solar activity. The changes that interfered with this periodicity he attributed to chaotic solar behavior.

Pisias (1978, 1979) reconstructed a time series curve of paleo-temperatures based on the analysis of *Radiolaria* assemblages in an annually laminated sediment core. Using this time series, a dataset reconstructed from all observed sea-surface temperatures and hydrographic data for the California Current, he developed a model to reconstruct sea-surface temperatures and dynamic height anomaly distributions for the California Current during the last 8000 years. This reconstruction indicated that the flow of the California Current was much stronger when sea-surface temperatures were low than it was during warm periods, like that of the present.

McKenzie and Eberli (1987) investigated the oxygen isotope stratigraphy of the sediments of the Great Salt Lake, Utah (Fig. 5.2). They found a good correlation between heavier values of the isotopes and the presence of aragonite. This could be correlated with periods of decreased freshwater inflow to the lake. Their reconstruction of the history of the Great Salt Lake correlated well with the history of paleo-environments of the northeastern Great Basin during the Holocene reconstructed by Currey and James (1982). According to this latter reconstruction, during the period from 7 ka to 5.5 ka BP, the Great Salt Lake was at a very low level, almost near a stage of complete desiccation. Wetter periods around 5 ka, 3.5 ka and 0.6 ka BP followed. These were cold periods of increased westerly flow and winter precipitation. In between, were warmer periods of low lake levels, associated with a strengthening of the Mexican monsoon activity. This happened at c. 4.5 ka and 1.5 ka BP.

By analyzing glacial deposits, Porter (1981) summarized the fluctuations of Holocene glaciers in western North America. In the curve he composed, one can observe four peaks of glacial advance before that of the Little Ice Age. These were c. 5 ka, 3 ka and 1 ka BP.

From the paleo-hydrological synthesis of Baker *et al.* (1995) it can be concluded that dry conditions existed during the Middle Holocene over the deserts of North America.

Pollen analysis from Wildcat Lake, Whitman County, Washington for the last 1000 years (Davis *et al.*, 1977) showed that terrestrial pollen percentages were relatively stable prior to the introduction of horses, sheep and cattle, but that changes occurred in the Wildcat Lake aquatic environment following deposition of volcanic ash from Mt. St. Helens (0.4 ka to 0.5 ka BP). Two periods of intense erosion followed the introduction of grazing and the ensuing range deterioration. It is obvious from these studies at Wildcat Lake that no climatic event of the past 1000 years resulted in vegetational changes as great as those brought about by European agricultural and grazing practices.

Stine (1998) focused on the period from c. 850 to 1325 AD. The first peak of cool and dry climate was during the first century since many lakes in western America dried up c. 850 AD. Stine suggested that this cold and dry spell was on a global scale as it correlated well with the advance of some Alaskan and Canadian glaciers. Another peak of cool and dry climate occurred for c. 50 years around 1130 AD. The general cold and dry conditions changed after 1325 AD, when the water levels of the desiccated water bodies started to rise again.

Ely *et al.* (1993) investigated the history of paleo-floods in Arizona and southern Utah. Their study was based on the mapping of fine deposits in backwater zones, preserved in protected niches outside the minimum flow bed of the river. The mapping of these deposits, along 19 riverbeds, enabled 251 extreme flood events to be distinguished, mainly clustered in the periods 5.2 ka to 3.6 ka, 2.2 ka to 0.8 ka and 0.6 ka BP to the present. From 3.6 ka to 2.2 ka BP, no sediments of large floods were recorded. A fall in the number of large floods occurred between 0.8 and 0.6 ka BP, immediately following a period of frequent large floods from 1 ka to 0.8 ka BP. According to these authors, the extended periods of large floods coincided with periods of cold climate, glacial advance and vegetation changes. By correlating the paleo-flood data with modern flood events and synoptic conditions, the authors found that large floods were connected to storms associated either with north Pacific winter fronts or with late summer and fall Pacific tropical cyclones, as well as with local convective summer storms. The two first conditions most probably caused (and would continue to cause) the largest floods. There was also a concurrence between large floods and El Niño events.

Correlating the climate changes of southwestern USA with those of the Levant and their impact on the hydrological cycle, one can see that during the warm and dry period of *c.* 4 ka BP in the Levant there is an increase in precipitation in southwest USA (Hughes and Graumlich, 1996). The same warming and increase of precipitation can be concluded from the data from Montezuma Well (Davis, 1995). The isotopic data from Great Salt Lake (McKenzie and Eberli, 1987) also indicated a warming trend.

The other isochrone most probably of global significance is that of the cold period starting *c.* 3.5 ka BP (Iron Age in the Levant, Jomon cold stage in Japan). This climatic trend was obvious in the two time series from the coast of California (Davis, 1995; Pisias, 1978). In Arizona, a general trend of less precipitation was observed (Hugues and Graumlich, 1996) and in the Great Salt Lake, one can see the starting of a trend towards depletion of oxygen isotopes (McKenzie and Eberli, 1987).

The cold period starting *c.* 2.3 ka BP and extending to 1.4 ka BP, with a short warm period from 1.7 ka to 1.6 ka BP (the Roman–Byzantine in the Levant and the Kofun period in Japan), can be observed in the two time series curves along the coast of California (Davis, 1992; Pisias, 1978). Lake Mono in eastern California was at a low level during this period (Stine, 1994). Analysis of palynological assemblage at Montezuma Well in Arizona (Davis, 1995) indicates lower temperatures in the lower altitude of the tree lines during this period. The tree line in the Lake Wright Basin (Lloyd and Graumlich, 1997) was also rather low while the ratio of spruce to pine in the southern Rocky Mountains (Petersen, 1994) was rather high. This coincided with a period of reduced precipitation on the Great Basin (Hughes and Graumlich, 1996). The ^{18}O isotope composition of the sediments of the Great Salt Lake was highly depleted, which implies minimum evaporation and thus a high lake level, and most probably a depleted composition of the precipitation to start with (McKenzie and Eberli, 1987).

Stine (1998) discussed at length the impact of the warm period from *c.* 1.3 ka to 1 ka BP and termed it the Medieval Climatic Anomaly (Arab period in the Mediterranean, the Nara–Heian warm stage in Japan, the Medieval Warm Period in Europe). The temperature of the sea surface off eastern California (Pisias, 1978) was mostly high. In the Great Salt Lake data, most of this period is characterized by a heavy isotope composition, indicating, in particular, high evaporation.

Comparing all these changes with variations in the width of the tree rings of Bristol Cone Pine growing near the upper tree line on the White Mountains in California (LaMarche, 1974) raises some doubt with regard to the conclusion of LaMarche that the width is an indication of variations of summer warmth and/or its seasonal duration. One gets a better correlation between the width of the tree rings and colder periods, most probably because of higher precipitation.

Some general conclusions can be drawn with regard to climatic changes and their impact on the hydrological cycle in the western USA during the Holocene. During cold periods, there was a southern shift of the westerlies zone, resulting in an abundance of winter rains and floods in the regions affected by this system. During warm periods, there was a shift northward of the summer rain monsoonal system and summer storms may have prevailed. However, these did not cause large floods, except during periods of El Niño events. This may be because the Colorado basin lies on the edge of the summer rain region and, therefore, storms that come from the Caribbean or Pacific tend to be moderate.

As for the past being a key to predict the future, one should also consider that the Colorado basin is on the border of two climate systems and is not far from the Pacific coast. Consequently, it is rather strongly influenced by El Niño events, making the prediction of its climate regime, and thus its hydrology, rather difficult. Although it can be generally foreseen that global warming would result in the summer storm belt moving northward, one cannot predict the extent of this shift. Overall, it can be said that a warm global change would cause higher summer precipitation rates and floods in the southern part of the Colorado basin, while in the northern part, the winter and summer rates of precipitation would be less. This scenario may change from time to time during El Niño events and/or volcanic events of large magnitude, which affect the transparency of the atmosphere on a global scale.

For California, it can be said that, in general, this region will see more summer rains coming from the Pacific, but winters will be drier. Again, during periods of El Niño events, torrential winter storms can be forecast.

5.2 CENTRAL USA AND CANADA

The impact of a warm peak around 11.6 ka BP on the hydrological cycle in North America can be established from the fact that meltwater, flooding down the Mississippi, reduced surface seawater salinities in the Gulf of Mexico by 10% (Emiliani, 1980; Roberts, 1989, p. 53).

Dean *et al.* (1996) showed that climatic warming to *c.* 2 °C above present values caused precipitation to be reduced by about 100 mm between 7.8 ka and 4.5 ka BP in North America, with a maximum reduction *c.* 6.8 ka BP (cal.). The dryness, which may have caused a decrease in vegetation cover, brought increased eolian activity. Varve calibrated records from Elk Lake and Lake Ann in Minnesota and showed that this eolian activity began at about 8 ka BP and ended at about 3.8 ka BP, with an interlude of reduced activity from 5.4 ka to 4.8 ka BP. The authors suggest that the increase in aridity and eolian activity was connected with increased solar geomagnetic activities.

Knox (1993) reported on the impact of climatic changes on floods in the upper Mississippi valley. Flooding episodes in this region are the result of excess rainfall during summer, when slow moving cold fronts arrived with moist and unstable tropically derived air masses. He equated periods of low probability of recurrence of high floods with warm and dry climates. Such floods occurred between 5 ka and 3.3 ka BP when, according to proxy-data, the mean July temperature was higher by about 0.5 °C and precipitation was about 15% less than modern values. After 3.3 ka BP, large floods became more frequent. The annual temperature decreased by about 0.7 °C and precipitation increased by about 8%. More frequent large floods can also be detected after 1 ka BP, especially between 0.7 ka and *c.* 0.5 ka BP.

Kay (1979) investigated the pollen assemblages in four core holes along a transection from the boreal forest limit into the tundra in the eastern Northwest Territories of Canada. He processed the data by multivariate statistical methods to reconstruct mean July paleo-temperatures. In the curves that he presented, one can see a rather warm period between 5.5 ka and 4 ka BP, with a short cooling phase *c.* 4.5 ka BP. At 3.7 ka BP, there was a major cooling phase, which extended to *c.* 2 ka BP. Another cooler phase occurred at 1.3 ka BP as well as at *c.* 0.5 ka BP.

There seems to be only minor correlation between Kay's (1979) climatic change records and the other data from North America discussed above, not to speak of the Levant. The question is whether the tools used in this investigation were responsive enough to climatic changes, a question that Kay himself raised.

Szeicz and Macdonald (1996) devised a ring-width chronology spanning 930 years for northwestern Canada, giving a correlation between moisture availability and global temperature in this region. In general, it can be concluded that a warmer climate brought more precipitation, and hence increased the width of rings, while a cold climate had the opposite effect. (This is different from the situation in the White Mountains of California (LaMarche, 1974), where wider rings were, I believe, associated with colder temperatures and higher precipitation, because of the Mediterranean-type climate that characterizes this region.)

Evison *et al.* (1996) investigated the history of the Late Holocene glaciation in the northeastern Brooks Range, in Alaska. The glaciers in this range are among the least active in the USA, because they are remote from major precipitation sources. At present they are declining. The major advances of these glaciers occurred at *c.* 2.6 ka, 1 ka, 0.45 ka and 60 BP (lichenometric years).

5.3 MEXICO

5.3.1 Contemporary climate

The contemporary climate of central and southern Mexico is influenced by two main atmospheric circulation systems: the trade winds and the sub-tropical high-pressure belt. The northern part of Mexico is also influenced by the westerlies. During summer, from April to October, the ITCZ moves north accompanied by the semi-permanent Bermuda–Azores and the East Pacific highs. This enables the trade winds to bring in moisture from the Caribbean as monsoonal rains. During August, there is a period of drier weather, the length of which decides the range of the annual average of precipitation. Between November and March, the ITCZ moves southward and the country comes under the influence of the sub-tropical high-pressure belt, with mainly a westerly wind direction. This causes stable dry weather. In the late winter and early spring, north and northwesterly depressions may bring precipitation and polar cold air to the northern part of the country, but in extreme cases they may bring rain as far as Yucatan. The multi-annual variations in the average quantity of precipitation has a marked influence on the levels of the lakes of Mexico, and one of the most important sources of data is that concerning the levels of the lakes of the Basin of Mexico (O'Hara and Metcalfe, 1997).

5.3.2 Climate changes during the Holocene

Metcalfe (1997) has studied the sediments of the lakes in the Zacapu Basin of central Mexico west to the Basin of Mexico City. No sediments were found for the period of transition from the Pleistocene to the Holocene. During the Holocene, only minor changes occurred in the levels of the lakes, except at *c.* 5.0 ka and 1.1 ka BP, when the lakes dried up.

Lozano-Garcia and Xelhuantzi-Lopez (1997) have studied the pollen, diatoms, macrofossil plant remains and tephras from cores in the sediments of this lake. They observed three main climatic stages. From *c.* 8.1 ka to 6.7 ka BP, a temperate sub-humid climate prevailed, which brought into existence a shallow lake with a low rate of sediment accumulation. From *c.* 6.7 ka to 5 ka BP, there was a reduction in humidity, and a temperate semi-arid climate prevailed. This caused the lake level to drop. During the last 5000 years, a temperate sub-humid climate prevailed.

Arnauld *et al.* (1997) made a synthesis of the various investigations of this basin and came up with the following division into Holocene phases.

Phase A: before *c.* 8 ka BP. The climate was humid to sub-humid and the lake was quite shallow.

Phase B: *c.* 8 ka to *c.* 7 ka BP. This was characterized by major volcanic activity, which interfered with the topography of the lake and the pollen record.

Phase C: *c.* 7 ka to *c.* 6 ka BP. In this period, the region stabilized and pollen and diatoms indicate a reduction in humidity with a very gradual change towards a semi-arid temperate climate.

Phase D: *c.* 6 ka to *c.* 4 ka BP. It was mainly dry and a shallow alkaline marsh existed. At about 4 ka BP, sub-humid climatic conditions returned and the marsh expanded again.

Phase E: *c.* 4 ka to *c.* 2 ka BP. There was a marked human impact on the environment. The climate became a little wetter and open water areas developed in the marsh.

The Aztecs settled to the east in the Basin of Mexico at *c.* 1345 AD, most probably driven from the north by particularly dry conditions, which were connected with a severe El Niño event (Manzanilla, 1997). After their settlement the climate became more humid, and at *c.* 1382 AD, the level of Lake Texcoco rose, flooding Tenochtitlan (now Mexico City) for 4 years. Extreme wet conditions prevailed until 1450 AD, after which a few years of extremely cold and dry conditions prevailed. During these dry years, which were also a period of famine, the temperatures were below normal levels. Later, and prior to the conquest by the Spaniards (around 1500 AD), the climate was very wet and the Aztecs had to build dams to protect Tenochtitlan from being flooded. A shift to slightly drier conditions, and droughts, occurred after the conquest by the Spaniards, during the period 1521–1640 AD, which suggests that the country was influenced by the sub-tropical pressure belt. Still during this period the climate was rather variable, and cold and dry years alternated with warm and humid periods. However, from *c.* 1640 to 1915 AD, the climate was dominated by droughts, which intensified during the mid to late 1700s, and late 1800s. Since 1915, there has been a shift towards wetter conditions. Therefore, it can be concluded that the period which coincided with the Little Ice Age was also a period of

droughts and famine. This was caused by the southward displacement of the ITCZ, suppressing the monsoons. During periods of El Niño, however, north Mexico may get higher rates of precipitation coming from the Pacific during winter, while the central part of Mexico still remains dry (O'Hara and Metcalfe, 1997).

Jauregui (1997) gave evidence for a strong causal link between cold periods, drought, famine and social unrest in Mexico. The climax was the drought years that preceded the 1810 and 1910 revolutions.

Hodell *et al.* (1995) have investigated the oxygen isotope composition and chemistry of a sediment core taken from the bottom of Lake Chichananab in the Yucatan peninsula. They found that the interval between 1.3 ka and 1.1 ka BP (800–1000 AD) was the driest of the Middle to Late Holocene and coincided with the collapse of classic Maya civilization. Whitmore *et al.* (1996) traced climatic and human influences on the nature of the sediments of three lakes. The climate impact started at *c.* 8 ka BP as groundwater levels rose as a result of a rise in sea level and increased precipitation. In its initial stages, Lake Coba, the most eastern lake and in the more humid part of the peninsula, was shallow and brackish but at *c.* 2.6 ka BP, the lake was at a high level and freshwater was present. This was followed by a decline in the water salinity and later an increase in salinity that continued to the present time. In Lake Sayaucil, in the intermediate precipitation zone, total salinity was high from 3.05 ka to 2 ka BP, followed by consistently higher lake levels. In San Jose Chulcaha, a lake in the most western and drier zone, salinity was high at 1.86 ka BP. This was followed by a gradual freshening of the water to the present time.

Gunn *et al.* (1995) have constructed a model that links global climate changes with the hydrology of the Candelaria river drainage basin and with agricultural production in the Yucatan peninsula. Highest levels of discharge and longest wet seasons were associated with global high temperatures, while lesser amounts of discharge and longer dry seasons were linked with low global temperatures. They came also to the conclusion that the general Maya collapse was caused by a severe drought period.

In northern Guatemala, variations in the pollen assemblage in a core in Lake Peten-Itza records the impact of climate as well as humans on the local environment. From 8.6 ka to 5.6 ka BP, the tropical forest dominated. The decline that followed was either a result of a climate change towards drying or caused by human impact. The clearance of the forest by the Mayan people and its substitution by agricultural plants was clearly evidenced beginning at about 2 ka BP (106 BC to 122 AD (cal.)). The forest returned after the Mayan collapse at *c.* 900 AD (Islebe *et al.*, 1996).

6 General conclusions

6.1 CLIMATE CHANGES DURING THE HOLOCENE: GLOBAL CORRELATION

In Fig. 6.1, representative time series from the regions discussed in the preceding chapters are correlated. From the correlation lines suggested in this figure, it can be concluded the main climate changes that occurred during the Holocene in the Levant can be traced in all other regions, although the range and nature of impact on the hydrological cycle differed from one region to another. As was discussed in Chapter 1, the archaeological stratigraphy in the Levant was by and large decided by the main climate changes. Moreover, archaeological investigations as well as historical documentation in this region are most extensive compared with other regions of the globe (Issar and Zohar, under revision). It is, therefore, suggested that the archaeological–paleo-climatic stratigraphy of the Levant should be adopted as the basic chronostratigraphy of the Holocene on a global scale.

The main chronostratigraphical divisions are listed in the key under Fig. 6.1 (from bottom to top) and shown on the diagram.

Some general conclusions can be drawn concerning the climate changes on a global scale demonstrated in Fig. 6.1.

1. During the Holocene, the global climate went through more than a few pronounced changes, which affected the hydrological cycle, the impacts of which were different from region to region on the time, temperature and humidity scales.
2. In the regions dominated by the westerlies, described earlier in this book (the circum-Mediterranean region, western and central Europe and western USA), cold climate brought more precipitation, which caused the hydrological systems to overflow. In the higher latitudes and altitudes, this cold climate caused glaciation. In these regions, warm climates, by comparison, led to less precipitation, causing desertification of the regions along the margins of the deserts.
3. In the more continental part of Europe (the drainage basin of the Caspian Sea), a cold climate brought dryness and vice versa.

4. In the monsoonal belts, cold periods were characterized by dryness. During these periods, these regions were influenced by the continental air masses with high barometric pressure, causing dust storms. Warm periods spelled more precipitation in the monsoonal regions. Floods, according to paleo-hydrological data, were on the increase.
5. In regions lying on the border zone between the westerlies and the monsoonal system, cold climates brought more winter rains and less summer rains and vice versa. The rate of dominance of each season depended on rate of shift of climate belts, which was mainly a shift of the change in global temperature.

6.2 THE CAUSE OF THE CLIMATE CHANGES DURING THE HOLOCENE

It is still debated exactly what mechanisms have caused the climate changes during the Holocene. Undoubtedly there are extraterrestrial, planetary, solar, atmospheric, hydrospheric and lithospheric factors, all exerting their influence on the climate on Earth, usually in a complex way that is not fully understood. Kutzbach and Street-Perrott (1985) suggested that the Milankovitch forcing mechanism has been largely responsible for the climatic changes affecting lake levels in the northern hemisphere tropics and sub-tropics since the Last Glacial Period. Their suggestion was based on the good correlation they achieved between an atmospheric general circulation model, simulating the climates of January and July at intervals of 3000 years, and the evidence regarding paleo-levels of lakes in the northern hemisphere tropics and sub-tropics. Both model and observations show high levels between 9 ka and 6 ka BP. However, as Kutzbach and Street-Perrott (1985) noted, the response of the monsoonal system to the amplified seasonal cycle of solar radiation, as a function of the variations in Earth's orbit, is a phenomenon that occurs on too long a time scale to account for even millennium-long, never mind century-long, climatic changes. Mörner (1984), Fairbridge (1984) and Ruddiman

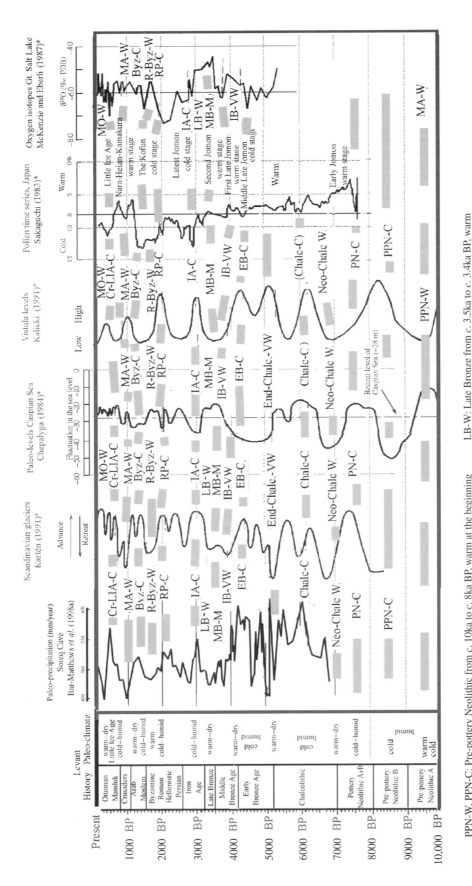

Fig. 6.1. Correlation of Holocene paleo-climates on a global scale. *Adjusted to scale and streamlined (3–5) points by the running average method.

PPN-W, PPN-C: Pre-pottery Neolithic from c. 10ka to c. 8ka BP, warm at the beginning cold towards the end

PN-C: Pottery Neolithic from c. 8ka to c. 7ka BP, mostly cold

Neo-Chalc W.: Neolithic–Chalcolithic transition, c. 7ka BP, warm

Chalc-C: Chalcolithic period c. 7.0ka to c. 5ka BP, cold

End-Chalc.-VW: end of Chalcolithic c. 5ka BP, short very warm interlude

EB-C: the Early Bronze Age c. 5ka to c. 4ka BP, cold

IB-VW: Intermediary Bronze Age c. 4ka BP, short very warm period

MB-M: Middle Bronze Age from c. 4ka to 3.5ka B.P, moderate climate

LB-W: Late Bronze from c. 3.5ka to c. 3.4ka BP, warm

IA-C: Iron Age from c. 3.4ka to c. 2.3ka BP, cold at the beginning later moderate

RP-C: Roman period from c. 2.3 ka to c. 1.7ka BP, cold

R-Byz-W: Roman–Byzantine intermediate period from c. 1.7ka to c. 1.6ka BP, warm

Byz-C: Byzantine period from c. 1.6ka to c. 1.3ka BP

MA-W: Moslem–Arab period from c. 1.3ka to c. 1ka BP, warm

Cr-LIA-C: Crusaders to Little Ice Age, from c. 1ka to 0.4ka BP, cold

MO-W: Moslem–Ottoman period, c. 0.4ka to 0.1ka BP, warm

Industrial period from c. 0.1ka BP, cool then turning to become warm and Duplessy (1985) maintained that, although the driving force behind a large fraction of Quaternary climatic changes lies in orbital variations, there are likely to be other phenomena that act as triggers of climate change, especially when they occur simultaneously with an orbital insolation signal, for example sun spots (Degens et al., 1984) or volcanic eruptions (Lamb, 1968). When it comes to changes during the Holocene, the latter most probably plays the major role.

Schove (1984) claimed that the relationship of sunspots and weather depends on the magnitude of the sunspot cycles. In the case of the medieval Nile floods, he found that the sign of relationship was reversed for strong and weak cycles.

Porter (1981, 1986) suggested that sulfur-rich aerosols, generated by volcanic activity, are a primary factor in forcing climate change on the decade level. Also Bryson and Bryson (1996, 1998, 2000) proposed that volcanic eruptions were the main triggering factor for climate changes during the Holocene. They suggested that explosions strong enough to load the atmosphere with a quantity of aerosols affected atmospheric optical depth and thus solar energy reaching the oceans. Taylor et al. (1993) measured the electrical conductivity of the layers of a core in the summit of the Greenland ice sheet. This is affected by the dust content of the layers, which acts to neutralize their acidity. Synchronicity was found between the ^{18}O composition and the electrical conductivity records, indicating that the latter is sensitive to climate changes. Layers with heavier ^{18}O composition, corresponding to warmer climates, also show higher electrical conductivity values: evidence of lower dust content. Scuderi (1990) has maintained that volcanic eruptions are the forcing mechanism causing climate changes on the time scale of tens to hundreds of years, leading to short colder periods; a cluster of significant events may lead to glaciation. Bradley et al. (1987) found that extremely sharp drops in temperature – particularly in autumn months – occurred after several major volcanic eruptions. High temperatures are sometimes associated with major El Niño years. When the two occur more or less simultaneously, their influence is minimized

Street-Perrott and Perrott (1990) found a connection between the droughts that have affected the Sahel and tropical Mexico since 14 ka BP and the injections of freshwater into the northern north Atlantic. This decrease in salinity causes a reduction in the formation of north Atlantic deep water circulation, which is an important factor in the regulation of the global oceanic circulation. This injection causes cooling of the sea surface and changes in the distribution of sea-surface temperatures, which affects rainfall in the northern tropics. The reasons for the freshwater injection phenomenon are not yet clear. On the basis of data obtained from a deep-sea sediment core, Chapman and Shackleton (2000) found evidence for cyclicity on a 550- and 1000-year cycle in north

Atlantic circulation patterns during the Holocene. Hughes et al. (2000) found a cyclicity of c. 600 and c. 1100 years between wetness periods in mires in Cumbria, England.

According to Dansgaard et al. (1984), the δ^{18}O oscillations in deep Greenland ice cores exhibit a quasi-periodicity of 2550 years. They concluded that the forcing of these cycles could be connected with cyclicity of solar radiation. Since the Milankovitch effects are too slow for this periodicity, they suggested that the reason might lie in solar productivity. They also pointed to evidence that ^{10}Be (a cosmogenic isotope) concentration in the ice is in an opposite phase with that of δ^{18}O. They, therefore, speculated that low solar activity, in terms of particle emissions, is connected with cold climates. However, Oeschger et al. (1984) suggested that variations in ^{10}Be concentrations mainly reflect changes in the precipitation rate.

Van Geel and Renssen (1998) suggested that the cold period between c. 850 and 760 BC was caused by reduced solar activity. Ji et al. (1993) claimed that spectral analysis of climate changes in China and aurora number series of the last 1700 years shows some potential relation at the time scales of decades to centuries. Tsoar (1995) observed desertification processes caused by desiccation in the northern Sinai desert during most of the seventeenth and eighteenth centuries AD. The dry period followed a wet period. Tsoar found some congruity between the wet period and the period of maximum sunspot activity known as the Medieval Maximum. In his opinion, the desiccation coincided approximately with the period of minimum solar activity known as the Maunder Minimum. When it comes to short-term fluctuations, recent investigations have shown that there is a link between the pattern of decadal changes in the temperature of the sea surface in the Pacific Ocean and the pattern of droughts and stream flow in the USA (Nigam et al., 1999).

Consequently, it is very difficult to argue against the opinion that volcanic activity should be regarded as an important, if not the major, cause for the periods of cold climates during the Holocene (Sadler and Grattman, 1999). However, the problem with volcanic forcing is that it causes only short-term events and, therefore, additional mechanisms must be sought to explain some of the long cold periods that lasted for more than a few centuries during the Holocene. I would suggest that a chain of events occurred, starting with strong volcanic activity. This caused short cold periods, which then caused high-pressure systems over the Gobi and Sahara deserts. These high-pressure systems led to heavy dust storms like those that characterized the Glacial and cold Post Glacial Periods, as well as historical periods (An Zhisheng et al., 1991a; Issar, 1990; Issar and Bruins, 1983; Lézine, 1998; Sarntheim et al., 1981; Taylor et al., 1993). These dust storms caused "red rains", such as those that take place contemporarily during cold spells and that also occurred during historical cold periods, when cyclonic storms entered the Sahara and moved over southern Europe (Bücher, 1986). These "red rains" are rich in iron oxide and phosphate and act as fertilizers on the bio-environment of the oceans. This causes a bloom of phyto-plankton, as a short-term CO_2 sink, which, in turn, causes a bloom of zoo-plankton to form a long-term sink, by sequestering CO_2 in the carbonate skeletons. This continues until a certain equilibrium is reached. Once there is a period of less volcanic activity, the CO_2 slowly returns to the atmosphere and a warm steady state, characterizing postglacial periods, is reached once again.

References

Aaby, B. (1976). Cyclic climatic variations in climate over the past 5500 yr reflected in raised bogs. *Nature*, 263, 281–284.

Abell, P. I. and Hoelzmann, P. (2000). Holocene paleoclimates in northwestern Sudan: stable isotope study on mollusks. *Global and Planetary Change*, 26, 1–12.

Adamson, D. A., Gasse, F., Street, F. A. and Williams, M. A. J. (1980). Late quaternary history of the Nile. *Nature*, 288, 50–55.

Aharoni, Y. (1967). The Israeli fortresses in the Negev. In *The Military History of the Land of Israel in the Biblical Times*, ed. Y. Liver, pp. 426–437. Tel Aviv: Israeli Defense Ministry Publishing.

Aharoni, Y. (1978a). The Israelite period. In *The Archaeology of the Land of Israel*, pp. 137–241. Jerusalem: Shikmona Publishing (in Hebrew).

Aharoni, Y. (1978b). *The Archaeology of the Land of Israel*. pp. 83–103. Jerusalem: Shikmona Publishing (in Hebrew).

Aharoni, Y. (1979a). The Negev in the Israelite period. In *The Land of the Negev – A Man and a Desert*, vol. 1, eds. A. Shmueli and Y. Grados, pp. 207–225. Tel Aviv: Israel Defense Ministry Publishing (in Hebrew).

Aharoni, Y. (1979b). *The Land of the Bible – A Historical Geography*, 2nd edn. London: Burns and Oates, 481 pp.

Aharoni, Y. (1986). The land of Israel in the Late Canaanite period and the Israelite settlement period. In *The History of the Land of Israel*, vol. 1, ed. Y. Ripel, pp. 87–178. Tel Aviv: Israeli Defense Ministry Publishing (in Hebrew).

Albright, W. F. (1926). From Jerusalem to Bagdad down the Euphrates. *Bulletin of the American Schools of Oriental Research*, 21, 1–21.

Albright, W. F. (1949). *The Archaeology of Palestine*, pp. 80–109. Hammondsworth, UK: Pelican-Penguin Books.

Albright, W. F. (1962). The chronology of Middle Bronze I (Early Bronze–Middle Bronze). *Bulletin of the American Schools of Oriental Research*, 168, 36–42.

Albright, W. F. (1965). Some remarks on the archaeological chronology of Palestine before about 1500 B.C. In *Chronologies in Old World Archaeology*, ed. R. W. Ehrich, pp. 47–60. Chicago, IL: University of Chicago Press.

Albright, W. F. (1973). The historical framework of Palestinian archaeology between 2100 and 1600 BC. *Bulletin of the American Schools of Oriental Research*, 209, 12–18.

Almogi-Labin, A., Hemleben, C. and Meischner, D. (1989). Palaeoenvironmental events in a Holocene sequence from the central Red Sea as recorded by pteropoda assemblages. In *Third International Conference on Paleooceanography, ICP III*, Cambridge 10–16 September, 1989, Abstract J1. Oxford: Blackwell Scientific.

Alon, D. (1988). The distribution and history of the Chalcolithic settlements in the southern Shefela region. In *Man and Environment in the Southern Shefela Region*, ed. D. Urman and A Stern, pp. 84–88. Beer Sheva: Ben Gurion University Press.

Alonso, A. and Garzon, G. (1994). Quaternary evolution of a meandering gravel bed in Central Spain. *Terra Nova*, 6, 465–475.

Alpert, P., Ben-Gai, T., Baharad, A. *et al.* (2002). Evidence for increase of extreme daily rainfall in the Mediterranean in spite of decrease in total values. *Geophysical Research Letters* 31, 1–4.

Altuna, J., Carreta, A., Edeso, L. M. *et al.* (1993). Yacimiento de Herriko-Bara (Zarautz, Pais Vasco). Y su Relacion con las Transgressiones Marinas Holoceanas., In *El Cuaternario en Espana Y Portugal, Acta de la 2 Reunion del Cuaternario Iberico* Madrid 25–29, September 1989, vol. 2, pp. 923–942.

Ambrosiani, B. (1984). Settlement expansion – settlement contraction: a question of war, plague ecology or climate? In *Climatic Changes on a Yearly to Millennial Basis*, eds. N. A. Mörner and W. Karlén, pp. 241–247. Dordrecht: Reidel.

Amiran, D. H. K. and Gilead, M. (1954). Early excessive rainfall and soil erosion in Israel. *Israel Exploration Journal*, 4, 286–295.

Amiran, R. (1985). The transition from the Chalcolithic to the EBA. In *Biblical Archaeology Today*, ed. J. Aviram, pp. 108–112. Jerusalem: Israel Exploration Society, Israel Academy of Sciences and Humanities and the American Schools of Oriental Research.

Amiran, R. (1986). The fall of the Early Bronze Age II city of Arad. *Israel Exploration Journal*, 36, 74–76.

Amiran, R. and Kochavi, M. (1985). The Land of Israel towards the end of the 3rd millennium B.C. – A separate culture or the last stage of Early Bronze Age? *Eretz Israel*, 18, 361–365 (in Hebrew).

Amiran, R., Amon, A., Alon, D., Githeart, R. and Lupen, P. (1980a). The old city of Arad, Results of 14 seasons of excavation. *Kadmoniot*, 19, 49–50 (in Hebrew).

Amiran, R., Alon, D., Amon, C. and Amiran, D. H. K. (1980b). The Arad countryside. *Levant*, XII, 22.

An Zhisheng, Wu Xihao, Wang Pinxian *et al.* (1991a). An evolution model for Paleomonsoon of China during the last 130 000 years In *Quaternary Geology and Environment in China, The Series of the XIII INQUA Congress*, ed. Liu Tungsheng, pp. 237–244. Beijing: Congress Science Press.

An Zhisheng, Wu Xihao, Wang Pinxian *et al.* (1991b). Paleoenvironmental changes of China during the last 18 000 years. In *Quaternary Geology and Environment in China, The Series of the XIII INQUA Congress*, ed. Liu Tungsheng, pp. 228–236. Beijing, China: Congress Science Press.

Anderson, D. E., Binney, H. A. and Smith, A. M. (1998). Evidence for abrupt climatic change in northern Scotland between 3900 and 3500 calendar years B.P. *The Holocene*, 8, 97–103.

Aoki, T. (1994). Chronological study of glacial advances based on the weathering ring thickness of morainic gravels in the northern part of the Central Japan Alps. *Geographical Review of Japan*, 67A, 601–618.

Araus, J. L., Febrero, A., Buxo, R. *et al.* (1997). Changes in carbon isotope discrimination in grain cereals from different regions of the western Mediterranean basin during the past seven millennia. Palaeoenvironmental evidence of a differential change in aridity during the late Holocene. *Global Change, Biology*, 3, 107–118.

ARIDUSEUROMED (1997). *Publication of Environment and Climate Program concerned with the Characterization of Aridity Processes on Mediterranean* Europe. Madrid: Library University of Complutense.

Arnauld, C., Metcalfe, S. E. and Petrequin P. (1997). Holocene climatic change in the Zacapu Lake Basin, Michoacan: synthesis of results. *Quaternary International*, 43/44, 173–179.

Ashbel, D. (1938). Great floods in Sinai Peninsula, Palestine, Syria and the Syrian desert and the influence of the Red Sea on their formation. *Quaternary Journal of the Royal Meteorological Society*, LXIV (277), 635–639.

Atkinson, T. C., Briffa, K. R. and Coope, G. R. (1987). Seasonal temperatures in Britain during the past 22,000 years reconstructed using beetle remains. *Nature*, 325, 587–592.

Avi-Yona, M. (1934). The Negev during the Byzantine period. *Yediot*, 2, 1–10, 44–50 (in Hebrew).

Avi-Yona, M. (1977). The Holy Land. From the Persian to the Arab conquests (536 BC to AD 564) – A historical geography. Grand Rapids, MC: Baker Book House.

Avner, U. (1998). Settlement, agriculture and paleoclimate in Uvda Valley, Southern Negev Desert, 6th–3rd millenia BC. In *Water, Environment and Society in Times of Climate Change*, eds. A. S. Issar and N. Brown, pp. 147–202. Dordrecht: Kluwer Academic.

Ayala-Carcedo, F. J. and Iglesiaz López, A. (1997). Impactos del Posible Cambio Climatico Sobre Los Recursos Hidricos, el Diseño y la Planificacion Hidrologica en la España Peninsular, Instituto Tecnológico Geominero de España (unpublished report). [They have based their investigations on the conclusions of the National Institute of Meteorology of Spain (Instituto Nacional del Metereologia, 1995, Informe de España sobre el Cambio Climático. Convenio Marco sobre el Cambio Climático de Naciones Unidas.]

Ayalon, A., Bar-Matthews., M. and Sass, E. (1998). Rainfall–recharge relationships within a karstic terrain in the eastern Mediterranean semi-arid region, Israel: ^{18}O and D characteristics. *Journal of Hydrology*, 207, 18–31.

Badal, E., Bernabeau, J., Dupre, M. and Fumanal, H. (1993). Secuencia Cultural y Paleoambiente en al Yacimiento Neolitico de la Cova de les Cendres (Moraira-Teulada. Alicante). In *El Cuaternario en Espana Y Portugal, Acta de la 2 Reunion del Cuaternario Iberico* Madrid, 25–29 September, 1989, vol. 2, pp. 943–953.

Barkay, G. (1992). The Iron Age II–III. *In The Archaeology of Ancient Israel*, ch. 9, ed. A. Ben-Tor, pp. 302–376. Tel-Aviv: The open University of Israel and New Haven, CT: Yale University Press.

Baker, V. R., Bowler, J.M., Enzel, Y., and Lancaster, N. (1995). Late Quaternary palaeohydrology of arid and semi-arid regions. In *Global Continental Palaeohydrology*, eds. K. J. Gregory, L. Starkel and V. R. Baker, pp. 203–231. Chichester, UK: John Wiley & Sons.

Bar-Matthews, M., Matthews, A. and Ayalon, A (1991). Environmental controls of speleothems mineralogy in a karstic dolomitic terrain (Soreq Cave, Israel). *Journal of Geology*, 99, 189–207.

Bar-Matthews, M., Ayalon, A., Matthews, A., Halicz, L. and Sass, E. (1993). The Soreq Cave speleothems as indicators of palaeoclimate variations. Geological survey of Israel. *Current Research*, 8, 1–3.

Bar-Matthews, M., Ayalon, A., Matthews, A., Sass, E. and Halicz, L. (1996). Carbon and oxygen isotope study of the active water–carbonate system in a karstic Mediterranean cave: implications for palaeoclimate research in semiarid regions. *Geochimica et Cosmochimica Acta*, 60, 337–347.

Bar-Matthews, M., Ayalon, A. and Kaufman, A. (1997). Late Quaternary paleoclimate in the eastern Mediterranean region from stable isotope analysis of speleothemes at Soreq Cave, Israel. *Quaternary Research*, 47, 155–168.

Bar-Matthews, M., Ayalon. A. and Kaufman, A. (1998a). Middle to Late Holocene (6,500 yr period) paleoclimate in the eastern Mediterranean region from stable isotopic composition of speleothemes from Soreq Cave, Israel. In *Water, Environment and Society in Times of Climate Change*, eds. A. Issar, and N. Brown, pp. 203–214. Dordrecht: Kluwer Academic.

Bar-Matthews, M., Ayalon, A., Kaufman, A. and Wassetburg, G.J. (1998b). The Eastern Mediterranean palaeoclimate as a reflection of regional events: Soreq Cave, Israel. *Earth and Planetary Science Letters*, 166, 85–95.

Barriendos, M. (1997). Climatic variations in the Iberian Peninsula during the late Maunder Minimum (AD 1675–1715): an analysis of data from rogation ceremonies. *The Holocene* 7, 105–111.

Bartov, Y., Stein, M., Enzel, Y., Agnon, A. and Reches, Z. (2002). Lake levels and sequence stratigraphy of Lake Lisan, the Late Pleistocene precursor of the Dead Sea. *Quaternary Research*, 57, 9–21.

Baruch, U. (1986). The Late Holocene vegetational history of Lake Kinneret (Sea of Galilee) Israel. *Paleorient*, 12, 37–48.

Bar Yosef, O. (1986a). The prehistoric period. In *The History of the Land of Israel*, vol. 1, ed. Y. Ripel, pp. 1–26. Tel-Aviv: Israeli Defense Ministry Publishing (in Hebrew).

Bar Yosef, O. (1986b). The walls of Jericho: an alternative interpretation. *Current Anthropology*, 27, 157–162.

Begin, Z. B., Broecker, W., Buchbinder, B. *et al.* (1985). Dead Sea and Lake Lisan levels in the last 30,000 years: A preliminary report. *Israel Geological Survey Report*, 29/85, 1–18.

Bell, B. (1971). The Dark Ages in ancient history. I. The first Dark Age in Egypt. *American Journal of Archaeology*, 75, 1–26.

Ben-Gai, T., Bitan, A., Manes, A., Alpert, P. and Rubin, S. (1998). Spatial and temporal changes in annual rainfall frequency distribution patterns in Israel. *Theoretical and Applied Climatology*, 61, 177–190.

Ben Tor, A. (1986). The Early Bronze Age. In *The History of the Land of Israel*, vol. 1, ed. Y. Ripel, pp. 61–72. Tel Aviv: Israeli Ministry Publishing (in Hebrew).

Ben Tor, A. (1992). The Early Bronze Age. In *The Archaeology of Ancient Israel*, ed. A. Ben-Tor, pp. 81–125. Tel Aviv: Open University of Israel; New Haven, CT: Yale University Press.

Berendsen, H. J. A. and Zagwijn, W.H. (1984). Geological changes in the western Netherlands during the period 1000–1300 AD. *Geologie en Mijnbouw*, 63, 225–229.

Bernier, P., Dalongeville, R., Dupuis, B. and de Medwecki, V. (1995). Holocene shoreline variations in the Persian Gulf: example of the Umm Al-Qowayn lagoon. *UAE Quaternary International*, 29/30, 95–103.

Bircher, W. (1986). Dendrochronology applied in mountain regions. In *Handbook of Holocene Paleoecology*, ed. B. E. Berglund, pp. 387–403. Chichester, UK: John Wiley & Sons.

Birks, H. J. B. (1986). Late-Quaternary biotic changes in terrestrial and lacustrine environments, with particular reference to north-west Europe. In *Handbook of Holocene Paleoecology and Paleohydrology*, ed. B.E. Berglund, pp. 3–66. Chichester, UK: John Wiley & Sons.

Björk , S. (1995). A review of the History of the Baltic Sea, 13.0–8.0 ka B.P. *Quaternary International*, 27, 19–40.

Bloch, M.R. (1976). Salt in human history. *Interdisciplinary Science Reviews*, 1, 336–352.

Blytt, A. (1876). *Essay on the Immigration of the Norwegian Flora During Alternating Rainy and Dry Periods*. Krisriania: Albert Cammeyer.

Bohncke, S. J. P. (1988). Vegetation and habitation history of the Callanish area, Isle of Lewis, Scotland. In *The Cultural Landscape – Past, Present and Future*, eds. H. H. Birks, H. J. B. Birks, P. E. Kaland and D. Moe, pp. 445–462. Cambridge, UK: Cambridge University Press.

Bohncke, S. J. P. (1991). Paleohydrological changes in the Netherlands during the last 13,000 years. PhD Thesis. Free University of Amsterdam, the Netherlands.

Bohncke, S. J. P and Vanderberghe, J. (1991). Paleohydrological development in the southern Netherlands during the last 15,000 years. In *Temperate Paleohydrology*, eds. L. Starkel, K.J. Gregory and J. B. Thornes. Chichester, UK: John Wiley & Sons.

Bonnefille, R. and Riollet, G. (1988). The Kashiru pollen sequence (burundi) paleoclimatic implications for the last 40,000 yr B.P. in tropical Africa. *Quaternary Research*, 30, 19–35.

Bonnefille, R., Roeland, J. C. and Guiot, J. (1990). Temperature and rainfall estimates for the past 40,000 years in equatorial Africa. *Nature*, 346, 347–349.

Bottema, S. (1978). Late glacial in eastern Mediterranean and the Near East. In *The Environmental History of the Near and Middle East Since the Last Ice Age*, ed. W.C. Brice, pp. 15–28. London: Academic Press.

Bowen, R. (1991). *Isotopes and Climates*, pp. 43–144. London: Elsevier Applied Science.

Bradley, R. S (1999). *International Geophysics Series:* vol. 64, *Paleoclimatology, Reconstructing Climates of the Quaternary*. San Diego, CA: Academic Press, 610 pp.

Bradley, R. S., Diaz, H. F., Jones, P. D. and Kelly, P. M. (1987). Secular fluctuations of temperature over northern hemisphere land areas and mainland China, since the mid-19th century. In *The Climate of China and Global Climate, Beijing International Symposium on Climate*, 30 October to November 1984, eds. Ye Dhuzheng, Fu Congbin, Chao Jiping and M. Moshino, pp. 76–87. Beijing: China Ocean Press and Springer Verlag.

Broshi, M. and Gophna, R. (1984). The settlements and population of Palestine during the Early Bronze Age II–III. *Bulletin of the American School of Oriental Research (BASOR)*, 253, 41–53.

Brown, N. (1998). Approaching the medieval optimum, 212 to 1000 AD. In *Water, Environment and Society in Times of Climate Change*, eds. A. Issar, and N. Brown, pp. 69–97, Dordrecht: Kluwer Academic.

Bruins, H. J. (1976). The origin, nature and stratigraphy of paleosols in the loessial deposits of the N. W. Negev. MSc Thesis, The Hebrew University, Jerusalem.

Bruins, H. J. and Yaalon, D. H. (1979). Stratigraphy of the Netivot section in the desert loess of the Negev (Israel). *Acta Geologica Academiae Scientarum Hungaricae*, 21, 161–169.

Bryson, R. A. and Bryson, R. U. (1996). High resolution simulations of regional Holocene climate: North Africa and Near East. In *NATO ASI Series, Subseries I Global Environmental Change: Climatic Change in the Third Millenium BC*, eds. N. H. Dalfes, G. Kukla and H. Weiss, pp. 565–593. Berlin: Springer Verlag.

Bryson, R. A. and Bryson, R. U. (1998). Application of a global volcanicity time-series on high resolution paleoclimatic modeling of the eastern Mediterranean. In *Water, Environment and Society in Times of Climate Change*, eds. A. Issar, and N. Brown, pp. 1–21. Dordrecht: Kluwer Academic.

Bryson, R. A. and Bryson, R. U. (2000). Site-specific high-resolution models of the monsoon for Africa and Asia. *Global and Planetary Change*, 26, 77–84.

Bücher, A. (1986). *Recherches sur les Poussieres Minerales d'Origine Saharienne*. PhD Thesis, University of Reims-Champagne, Ardenne.

Burney, D. A., Brook, G. A. and Cowart, J. B. (1994). A Holocene pollen record for the Kalahari Desert of Botswana from a U-series dated speleothem. *The Holocene* 4, 225–232.

Butzer, K. W. (1958). *Quaternary Stratigraphy and Climate in the Near East*, pp. 116–118. Bonn: Dummler.

Butzer, K. W. (1966). Climate changes in the arid zones of Africa during Early and Mid Holocene times. In *World Climates from 8000 to 0 B.C.*, pp. 72–83, London: Royal Meteorological Society.

Butzer, K. W. (1978). The late prehistoric environmental history of the Near East. In *The Environmental History of the Near and Middle East since the Last Ice Age*, ed. W.C. Brice, pp. 5–12. London: Academic Press.

Butzer, K. W. (1980). Pleistocene history of the Nile Valley in Egypt and Lower Nubia. In *The Sahara and the Nile*, ed. M. A. J. Williams and H. Faure, pp. 253–280. Rotterdam: Balkema.

Butzer, K. W., Isaac, G. L., Richardson, J. L. and Washbourn Kamu, C. (1972). Radiocarbon dating of East African Lake levels. *Science*, 175, 1069–1076.

Capel Molina, J. J. (1981). *Los Climas de España, Oikus-tau*. Barcelona: SA editions, 429 pp.

Carpenter, R. (1966). *Discontinuity in Greek Civilization*. Cambridge, UK: Cambridge University Press, 79 pp.

Carrión, J. S. and Dupré M. (1996). Late Quaternary vegetational history at Navarrés, Eastern Spain. A two core approach. *New Phytologist*, 134, 177–191.

Carrión, J. S., Andrade, A., Bennett, K. D., Navarro, C. and Munera, M. (2001). Crossing forest thresholds: inertia and collapse in a Holocene sequence from south-central Spain. *The Holocene*, 11, 635–653.

Casparie, W. A. (1972). Bog development in southeastern Drenthe (the Netherlands). PhD Thesis, University of Groningen, Published by Dr W. Junk N.V.s Gravenhage, 271 pp.

Chambers, F.M., Barber, K.E., Maddy, B. and Brew, J. (1997). A 5500-year proxy climate and vegetation record from blanket mire at Talla Mos, Borders, Scotland. *The Holocene*, 7, 391–399.

Chapman, M. R. and Shackleton, N. L. (2000). Evidence of 550-year and 1000-year cyclicities in North Atlantic circulation patterns during the Holocene. *The Holocene*, 10, 287–291.

Chepalyga, A. L. (1984). Inland Sea Basins. In *Late Quaternary Environments of the Soviet Union*, eds. A. A. Velichkoe, H. E. Wright Jr and C. W. Barnosky (English language edition), pp. 229–247, London: Longman.

Chernavaskaya, M. M. (1990). Climate of Eastern Europe in the historical past based on proxy data. In *Climatic Changes and Their Impacts, Beijing International Symposium on Climatic Change*, 9–12 August, p. A-9. Beijing, China: Natural Science Foundation of China.

Clutton-Brock, J. (1978). Early domestication and the ungulate fauna of the Levant during the Prepottery Neolithic Period. In *The Environmental History of the Near and Middle East Since the Last Ice Age*, ed. W. C. Brice, pp. 29–40. London: Academic Press.

Cohen, A. L., Parkington, J. E., Brundrit, G. B. and van der Merwe, N. J. (1992). A Holocene marine climate record in mollusc shells from the Southwest African Coast. *Quaternary Research*, 38, 379–375.

Cohen, R. (1979a). The settlements in the Negev mountain from the 4th millennium BC till the 4th century BC. *Kadmoniot*, 83–84, 62–81 (in Hebrew).

Cohen, R. (1979b). The archaeological survey in the Negev in the last years. *Kadmoniot* 45, 34–36 (in Hebrew).

Cohen, R. (1983). The mysterious MBI people. *Biblical Archaeology Review*, IX, 16–29.

Cohen, R. (1985). *An Archaeological Survey of Israel. Sede Boker-East Map (168)*. Jerusalem: Antiquity Division Publishing (in Hebrew).

Cohen, R. (1986). The settlements in Negev Highlands, according to the archaeological data, in the 2nd and 1st millennium, BC. PhD Thesis, The Hebrew University, Jerusalem.

Cohen, R. (1989). The settlements in the Negev mountain from the 4th millennium B.C. till the 4th century BC. *Kadmoniot*, 83–84, 62–81 (in Hebrew).

Coope, G. R. (1975). Climatic fluctuations in northwest Europe since the last interglacial, indicated by fossil assemblages of coleoptera. In *Ice Ages: Ancient and Modern*, eds. A. E. Wright and F. Moseley, pp. 153–168. Liverpool: Seel House Press.

Coope, G. R. (1987). Fossil beetle assemblages as evidence for sudden and intense climatic changes in the British Isles during the last 45,000 years. In *Abrupt Climatic Change*, eds. W. H. Berger and L. D. Labeyrie, pp. 147–150. Dordrecht: Reidel.

Crown, A. D. (1972). Toward a reconstruction of the climate of Palestine 8000 B.C.–0 B.C. *Journal of Near Eastern Studies*, 31, 312–330.

Cui Zhijiu and Song Changqing (1991). Investigated Quaternary periglacial environment in China. In *Quaternary Geology and Environment in China, The Series of the XIII INQUA Congress*, ed. Liu Tungsheng, 1991, pp. 78–85. Beijing, China: Congress Science Press.

Currey, D. R. and James, S. R. (1982). Paleoenvironments of the northeastern Great Basin and northwestern basin rim region: a review of geological and biological evidence. In *Man and Environment in the Great Basin*, ed. D. B. Madsen and J. F. O'Conell, SAA Paper 2, pp. 27–52. Washington, DC: Society for American Archaeology.

Dahl, S. O. and Nesje, A. (1996). A new approach to calculating Holocene winter precipitation by combining glacier equilibrium-line altitudes and pine-tree limits: a case study from Hardangerjøkulen central southern Norway. *The Holocene*, 6, 392–398.

Dalfes, N., Kukla, G. and Weiss, H. (eds.) (1997). *NATO ASI Series*, vol. 149: *Third Millenium BC Climate Change and Old World Collapse*. Berlin: Springer Verlag, 728 pp.

Dansgaard, W., Johnsen, S. J., Clausen, H. B. and Langway, C. C. (1971). Climatic record revealed by the Camp Century ice core. In *The Late Cenozoic Glacial Ages*, ed. K.K. Turekian, pp. 37–56. New Haven, CT: Yale University Press.

Dansgaard, W., Johnsen, S. J., Clausen, D. *et al.* (1984). North Atlantic oscillations revealed by deep Greenland ice cores. In *Climate Processes and Climate Sensitivity*, eds. J. E. Hansen and T. Takahashi, pp. 288–298. Washington, DC: American Geophysical Union.

Darmon, F. (1988). Essai de reconstitution climatique de l'epipaleolithique an debut du Néolithique ancien dans la region de Fazael-Salibiya (Basse Vallée du Jourdain) d'après la palynologie. *Comptes Rendus Academy des Sciences* (Paris). Serie II, 307, 677–682.

Davis, O. K. (1992). Rapid climatic change in coastal southern California inferred from pollen analysis of San Joaquim Marsh. *Quaternary Research*, 37, 89–100.

Davis, O. K. (1994). The correlation of summer precipitation in the southwestern USA with isotopic records of solar activity during the Medieval Warm Period. In *The Medieval Warm Period*, eds. M.K. Hughes and H.F. Diaz, pp. 271–287. Dordrecht: Kluwer Academic.

Davis, O. K. (1995). Climate and vegetation patterns in surface samples from arid western USA: application to Holocene climatic reconstructions. *Palynology*, 19, 95–117.

Davis, O. K. (1996). The impact of climate change on available moisture in arid lands, examples from the American southwest. In *NATO ASI Series, Subseries I*, vol. 36, *Diachronic Climatic Impact on Water Resources*, eds. A. N. Angelakis and A. S. Issar, pp. 283–300. Berlin: Springer Verlag.

Davis, O. K., Kolva, D. A. and Mehringer, P. J. Jr (1977). Pollen analysis of Wildcat Lake, Whitman County, Washington: the last 1000 years. *Northwest Science*, 51, 13–30.

Deacon, H. J., Deacon, J., Scholtz, A., Thackeray, J. F., Brink, J. S. and Vogel, J.C. (1984). Correlation of paleoenvironmental data from the Late Pleistocene and Holocene deposits at Boomplaas cave, S. Cape. In *Late Cainozoic Paleoclimates of the Southern Hemisphere*, ed. J.C. Vogel, pp. 339–353. Rotterdam: Balkema.

Deacon, J. and Lancaster, N. (1988). *Late Quaternary Paleo-Environments of Southern Africa*. Oxford: Clarendon Press.

Dean, W. E., Ahlbrandt, T. S., Anderson, Y. and Platt Bradbury, J. (1996). Regional aridity in North America during the middle Holocene. *The Holocene* 6, 145–155.

Degens, E. T. (1971). Sedimentological history of the Black Sea. In *Geology and History of Turkey*, ed. A. S. Campbell, pp. 407–429. Tripoli, Libya: Petroleum Exploratory Society of Libya.

Degens, E. T., Wong, H. K., Kempe, S. and Kurtman, F. (1984). A geological study of Lake Van, Eastern Turkey. *Geologische Rundschau*, 73, 701–734.

de Vaux, R. (1971). Palestine in the Early Bronze Age. In *The Cambridge Ancient History of the Middle East*, vol. 1(2), eds. I. E. S. Edwards, C. J. Gadd and N. G. L. Hammond, pp. 208–237. Cambridge: Cambridge University Press.

Dever, W. G. (1973). The EB IV–MB I horizon in Transjordan and southern Palestine. *Bulletin of American Schools of Oriental Research*, 210, 37–63.

Dever, W. G. (1976). The beginning of the Middle Bronze Age in Syria–Palestine. In *Magnalia Dei: The mighty Acts of Gods*, [*Essays on the Bible and Archaeology in Memory of G. Ernest Wright*], eds. F. M. Cross, W. E. Lemke and P. D. Miller, pp. 3–38. Garden City, NY: Doubleday.

Dever, W. G. (1980). New vistas on the 'EB IV–MB I' horizon in Syria–Palestine. *Bulletin of the American Schools of Oriental Research*, 237, 31–59.

Dever, W. G. (1985a). From the end of the Early Bronze Age to the beginning of the Middle Bronze. In *Biblical Archaeology Today*, ed. J. Aviram, pp. 113–135. Jerusalem: Israel Exploration Society, Israel Academy of Sciences and Humanities and the American Schools of Oriental Research.

Dever, W. G. (1985b). Relation between Syria–Palestine and Egypt in the "Hyksos" period. In *Palestine in the Bronze and Iron Ages* [*Papers in honour of Olga Tufnell*], ed. J. N. Tubb, pp. 69–87. London: Institute of Archaeology.

Dever, W. G. (1987). The middle bronze Age. The zenith of the urban Canaanite era. *Biblical Archaeology*, 50, 149–177.

Digerfeldt, G. (1988). Reconstruction and regional correlation of Holocene lake-level fluctuations in Lake Bysjö, South Sweden, *Boreas*, 17, 165–182.

Digerfeldt, G., de Bealieu, J.-L., Guiot, J. and Mouthon, J. (1997). Reconstruction of Holocene lake-levels changes in Lac de Saint-Léger, Hute-Provence, southeast France. *Palaeogeography, Palaeoclimatology and Palaeoecology*, 136, 231–258.

Dragoni, W. (1998). Some considerations on climatic changes, water resources and water needs in the Italian region south of 43 °N. In *Water, Environment and Society in Times of Climate Change*, eds. A. Issar and N. Brown, pp. 241–272. Dordrecht: Kluwer Academic.

Dupont, L. M. (1985). Temperature and rainfall variation in a raised bog ecosystem, a paleocological and isotope–geological study. PhD Thesis, University of Amsterdam.

Dupre, M., Fumanal, M. P., Martinez Gallego, J., Perez Obiol, R., Roure, J. O. and Usera, J. (1996). The laguna de San Benito, Paleoenvironmental reconstruction of an endorheic system. *Quaternaire*, 7, 177–186.

Elliot, C. (1978). The Chalcolithic culture in Palestine: origins, influences and abandonment. *Levant*, 10, 37–54.

El-Fishawi, N. M. and Fanos, A. M. (1989). *MBSS Newsletter* No. 11: *Prediction of Sea Level Rise by 2,100, Nile Delta Coast*, eds. C. Zaro and T. Bardaji, pp. 43–47. CINQUA, Commission on Quaternary shorelines, subcommission on Mediterranean and Black Sea shorelines. Madrid: University of Complutense.

El-Moslimany, A. P. (1986). Ecology and late quaternary history of the Kurdo Zagrosian oak forest near Lake Zeribar, western Iran. *Vegetation*, 68, 55–63.

Ely, L. L., Enzel, Y., Baker, V. R. and Cayan, D. R. (1993). A 5000-year record of extreme floods and climate change in the southwestern United States. *Science*, 262, 410–412.

Emery, K. O. and Neev, D. (1960). Mediterranean beaches of Israel. *Geological Survey of Israel Bulletin*, 26, 1–23.

Emiliani, C. (1955). Pleistocene temperature. *Journal of Geology*, 63, 538–578.

Emiliani, C. (1980). Ice sheets and ice melts. *Natural History*, 89, 82–91.

Erinc, S. (1978). Changes in the physical environment in Turkey since the end of the Last Glacial. In *The Environmental History of the Near and Middle East since the Last Ice Age*, ed. W.C. Brice, pp. 87–110. London: Academic Press.

Evenari, M., Shannan, L. and Tadmor, N. (1971). *The Negev: The Challenge of a Desert*. Boston, MA: Harvard University Press, 345 pp.

Evison, L. H., Calkin, E. and Elis, J. M. (1996). Late Holocene glaciation and twentieth century retreat, northeastern Brooks Range, Alaska. *The Holocene*, 6, 17–24.

Fairbridge, R. W. (1962). New radiocarbon dates of Nile sediments. *Nature*, 196, 108–110.

Fairbridge, R. W. (1984). Planetary periodicites and terresterial climate stress. In *Climatic Changes on a Yearly to Millennial Basis*, eds. N. A. Mörner and W. Karlén, pp. 509–520. Dordrecht: Reidel.

Fan Jianhua and Shi Yafeng (1990). The impact of climate change on the hydrologic regime of inland lakes: A case study of Qinghai Lake. In *Climatic Changes and Their Impacts, Beijing International Symposium on Climatic Change*, 9–12, August, p. C-24. Beijing: Natural Science Foundation of China (CNSF).

Fang Jin-Qix (1990). The impact of climatic change on the Chinese migrations in historical times. In *Regional Conference on Asian Pacific Countries of I.G.U.: Global Change and Environmental Evolution in China*, Section III, eds. Liu Chuang, Zhao Songqiao, Zhang Peiyuan and Shi Peijun, pp. 96–103. Hohot, P. R. China: Editorial Board of Arid Land Resources.

Faure, H. and Faure-Denard, L. (1998). Sahara environmental changes during the Quaternary and their possible effect on carbon storage. In *Water, Environment and Society in Times of Climate Change*, eds. A. Issar and N. Brown, pp. 319–322, Dordrecht: Kluwer Academic.

Ferronsky, V. I. and Polyakov, V. A. (1982). *Environmental Isotopes in Hydrosphere*. Chichester, UK: John Wiley & Sons, 466 pp.

Finkelstein, Y. and Perevolotsky, A. (1989). Settlement and nomadism in the deserts of the south in ancient periods. *Katedra*, 52, 3–39 (in Hebrew).

Finley, M. I. (1981). *Early Greece, the Bronze and Archaic Ages*. New York: Norton, 149 pp.

Finney, B. P. and Johnson, T. C. (1991). Sedimentation in Lake Malawi (East Africa) during the past 10,000 years: a continuous paleoclimatic record from the southern tropics. *Palaeogeography, Palaeoclimatology and Palaeoecology*, 85, 351–366.

Flohn, H. and Nicholson, S. (1980). Climatic fluctuations in the arid belt of the "Old World" since the last glacial maximum; possible causes and future implications. *Palaeoecology of Africa*, 12, 3–21.

Foucault, A. and Stanley, D. J. (1989). Late Quaternary paleoclimatic oscillations in East Africa recorded by heavy minerals in the Nile delta. *Nature*, 339, 44–46.

Fritts, H. C. and Shao, X. M. (1992). Mapping climate using tree-rings from western North America. In *Climate since A.D. 1500*, eds. R. S. Bradely and P. D. Jones, pp. 269–295. London: Rutledge.

Fritz, P. and Fontes, J. Ch. (1980). *Handbook of Environmental Isotope Geochemistry*, vol. 1. Amsterdam: Elsevier.

Frumkin, A., Magaritz, M., Carmi, I. and Zak, I. (1991). The Holocene climatic record of the salt caves of Mount Sedom, Israel. *The Holocene*, 1, 191–200.

Frumkin, A., Carmi, I., Gopher, A., Ford, D. C., Schwarcz, H. P. and Tsuk, T. (1999). A Holocene millennial-scale climatic cycle from a speleothem in Nahal Qanah Cave, Israel. *The Holocene*, 9, 677–682.

Fu-Bau, W. and Fan, C. Y. (1987). Climatic changes in the Qinghai-Xizang (Tibetan) region of China during the Holocene. *Quaternary Research*, 28, 50–60.

Fu Congbin, Dong Dongfeng and Wang Qiang (1990). A dry trend in past 100 years and the abrupt feature in its evolution. In *Climatic Changes and Their Impacts, Beijing International Symposium on Climatic Change*, 9–12 August, p. c-2. Beijing: Natural Science Foundation of China (CNSF).

Galili, E., Weinstein-Evron, M. and Ronen, A. (1988). Holocene sea-level changes based on submerged archaeological sites off the northern Carmel coast in Israel. *Quaternary Research*, 29, 36–42.

Garcia Anton, M., Moria Juraisti, C., Ruiz Zapata, B. and Sainz Ollero, H. (1986). Contribucion al conociento el paisaje vegetal Holoceno en la submeseta Sur Iberia: analisis polinico de sedimentos higroturbosos en el campo del Calarava (Ciudad Real) Espana. In *Proceedings of the Symposium of Climatic Fluctuations during the Quaternary in the Western Mediterranean Regions*, ed. F. Lopez-Vera, pp. 189–204. Madrid: Universidad Autonoma de Madrid.

Gasse, F. (1980). Late Quaternary changes in lake-levels and diatom assemblages on the southeastern margin of the Sahara. *Paleoecology of Africa*, 12, 333–350.

Gasse, F., Rognon, P. and Street, F. A. (1980). Quaternary history of the Afar and Ethiopian Rift lakes. In *The Sahara and the Nile*, eds. M. A. J. Williams and H. Fauke, pp. 361–400. Rotterdam: A. A. Balkema.

Gat, J. (1981). *Stable Isotope Hydrology, Deuterium and Oxygen-18 in the Water Cycle: A Monograph*. Vienna: IAEA, 339 pp.

Gat, J. R. and Issar, A. (1974). Desert isotope hydrology: water sources of the Sinai Desert. *Geochimica et Cosmochimica Acta*, 38, 1117–1131.

Gear, A. J. and Huntley, B. (1991). Rapid changes in the range limits of Scots Pine 4000 years ago, *Science*, 251, 544–546

Geyh, M. A. (1994). The paleohydrology of the eastern Mediterranean. In *Radiocarbon 1994: Late Quaternary Chronology of the Eastern Mediterranean*, eds. O. Bar-Yosef and R. S. Kra, pp. 131–145. Phoenix, AZ: Dept Geosciences, University of Arizona.

Geyh, M. A. and Franke H. W. (1970). Zur Wachtumsgeschwindigkeit von stalagmiten, *Atompraxis*. 16, 1–3.

Geyh, M. A. and Jäkel, D. (1974). Late glacial and Holocene climatic history of the Sahara desert derived from a statistical assay of ^{14}C dates. *Palaeogeography, Palaeoclimatology and Palaeoecology*, 15, 205–208.

Gillepsie, R., Street-Perrott, A. F. and Switsur, R. (1983). Post-glacial arid episodes in Ethiopia have implications for climate predictions. *Nature*, 306, 680–683.

Giraudi, C. (1989). Lake levels and climate for the last 30,000 years in the Fucino area. *Palaeogeography, Palaeoclimatology and Palaeoecology*, 70, 249–260.

Goldberg, P. (1977). Late quaternary stratigraphy of Gebel Maghara. *Qedem*, 7, 11–31.

Goldberg, P. and Rosen, A. M. (1987). Early Holocene palaeoenvironments of Israel. In *British Archeology Reports International Series 356(i): Shikmim I. Studies Concerning 4th Millennium Societies in the Northern Negev Desert, Israel*, ed. T.E. Levy, pp. 23–34. Oxford: British Archaeology Reports.

Gonen, R. (1992). The Chalcolithic period. In *The Archaeology of Ancient Israel*, ed. A. Ben-Tor, pp. 40–80. Tel Aviv: Open University of Israel; New Haven, CT: Yale University Press.

Goodfriend, G. A. (1990). Rainfall in the Negev desert during the Middle Holocene, based on ^{13}C of organic matter in land snail shells. *Quaternary Research*, 34, 186–197.

Goodfriend, G. A. (1991). Holocene trends in ^{18}O in land snail shells from the Negev Desert and their implications for changes in rainfall source areas. *Quaternary Research*, 35, 417–426.

Goodfriend, G. A. and Magaritz, M. (1988). Palaeosols and late pleistocene rainfall fluctuation in the Negev Desert. *Nature*, 332, 144–146.

Gopher, A. (1981). *The prehistory of the Central Negev*. Sede Boker, Negev, Israel: The Society for Nature Protection, Sede Boker Field School (in Hebrew).

Gophna, R. (1983). The Chalcolithic period. In *The History of the Land of Israel, the Ancient Periods*, ed. I. Efal, pp. 76–94. Jerusalem: Keter (in Hebrew).

Gophna, R. (1992). The Intermediate Bronze period. In *The Archaeology of Ancient Israel*, ed. A. Ben-Tor,. pp. 126–158. Tel Aviv: Open University of Israel; New Haven, CT: Yale University Press.

Goring-Morris, A. N. and Goldberg, P. (1990). Late Quaternary dune incursions in the Southern Levant: archaeology, chronology and paleoenvironments. *Quaternary International*, 5, 115–137.

Govrin, Y. (1991). The question of the origin of early Arad's population. In *The Land of Israel. Researches in Geography and Archaeology*, Vol 21, pp. 107–110. Jerusalem: Ruth Amiran Book. The Society for Exploring the Land of Israel and its Antiquities Publishing (in Hebrew).

Goy, J. L., Zazo, C., Dabrio, C. J. *et al.* (1996). Global and regional factors controlling changes of coastlines in southern Iberia (Spain) during the Holocene. *Quaternary Science Review*, 15, 773–780.

Goytre, M. J. and Garzon, G. (1996). Analisis de las Avenidas Historicas en el Rio Jucar. In *Proceedings of the VI Congress and International Conference of Geologia Ambiental y Ordenacion del Territoria*, 22–25 April, pp. 29–41.

Graumlich, L. J. (1993). A 1000-Year record of temperature and precipitation in the Sierra Nevada. *Quaternary Research*, 39, 249–255.

Graumlich, L. J. and Lloyd, A. H. (1996). Dendroclimatic, ecological and geomorphological evidence for long-term climatic change in the Sierra Nevada, USA. In *Radiocarbon 1996: Tree Rings, Environment and Humanity*, eds. J. S. Dean, D. M. Meko and T. W. Swetnam, pp. 51–59. Phoenix, AZ: Dept Geosciences, University of Arizona.

Griffiths, H. I., Schwalb, A., Stevens, L. L. R. (2001). Environmental change in southwestern Iran: the Holocene ostracod fauna of Lake Mirabad. *The Holocene*, 11, 757–764.

Guarong, D. (1988). The problem of formation, evolution and origin of Mauwusu Desert. *Scientia Sinica* (Series B), 6, 633–642.

Gunn, J. D., Folan, W. J., Robichaux, H. R. (1995). A landscape analysis of the Calendaria watershed in Mexico: insight into paleoclimates affecting upland horticulture in the southern Yucatan Peninsula Semi-Karst. *Geoarchaeology*, 10, 3–42.

Guo Qiyun (1990). An analysis of the teleconnection between the floods and droughts in North China and Indian summer monsoon. *Climatic Changes and Their Impacts, Beijing International Symposium on Climatic Change*, 9–12 August, pp. A-3, A-2. Beijing: Natural Science Foundation of China (CNSF).

Haas, J. N., Richoz, I., Tinner, W. and Wick, L. (1997). Synchronous Holocene climatic oscillations recorded on the Swiss Plateau and at timberline in the Alps. *The Holocene*, 8, 301–309.

Hageman, B. P. (1969). Development of the western part of the Netherlands during the Holocene. *Geologica Mijnbouw*, 48, 373–388.

Han Shu-ti and Yuan Yu-Jiang (1990). Research on the climate features and changing tendencies in Lake Barkol of Xingjiang since 35000 Yrs. In *Regional Conference on Asian Pacific Countries of I.G.U.: Global Change and Environmental Evolution in China*, Section III, eds. Liu Chuang, Zhao Songqiao, Zhang Peiyuan, Shi Peijun, pp.79–66. Hohot, P. R. China: Editorial Board of Arid Land Resources.

Hanyong, H. and Shanyu, Z. (1984). *Chinese Population, Geography*, Vol. 1, pp. 66–67, 325–342. China: East China Normal University Press.

Harlan, J. R. (1982). The garden of the Lord: a plausible reconstruction of natural resources of Southern Jordan in Early Bronze Age. *Paléorient*, 8, 71–78.

Harrison, S. P., Prentice, C. and Guiot, J. (1993). Climatic controls on Holocene lake level changes in Europe. *Climate Dynamics*, 8, 189–200.

Hastenrath, S. and Kutzbach, J. E. (1983). Paleoclimate estimates from water and energy budgets of East African lakes. *Quaternary Research*, 19, 141–153.

Hatano, S. (1979). Mapping of post-glacial dissected hillslopes and its application to landslide prediction. In *Proceedings of the Japanese Erosion-Control Engineering Society*, pp. 16–17.

Heaton, T. H., Talma, A. S. and Vogel, J. C. (1986). Dissolved gas paleotemperatures and ^{18}O variations derived from groundwater near Uitenhag, South Africa. *Quaternary Research*, 25, 79–88.

Heine, K. and Geyh, M. A. (1983). Radiocarbon dating of speleothems from the Rossing Cave, Namibia Desert and paleoclimatic implications. *In International Symposium of the South Africa Society for Quaternary Research*, Swaziland, pp. 465–470. Rotterdam: Balkema.

Hennessy, J. B. (1982). Teleilat Ghassul: its place in the archaeology of Jordan. In *Studies in the History and Archaeology of Jordan*, vol. I, ed. A. Hadidi, pp. 55–58. Amman: Department of Antiquities.

Herman, Y. (1989). Late quaternary paleoceanography of the Eastern Mediterranean. The deep sea record. *Marine Geology*, 87, 1–4.

Hodell, D. A., Curtis, J. H. and Brenner, M. (1995). Possible role of climate in the collapse of classic Maya civilization, *Nature*, 375, 391–394.

Hoelzmann, P., Kruse, H.-J. and Rottinger, F. (2000). Precipitation estimates for the eastern Saharan paleomonsoon based on a water balance model of the West Nubian Paleolake Basin. *Global and Planetary Change*, 26, 105–120.

Holmgren, K., Karlen, W., Lauritzen, S. E. *et al.* (1999). A 3000-year high-resolution stalagmite-based record of palaeoclimate for northeastern South Africa. *The Holocene*, 9, 295–309.

Horowitz, A. (1973). Development of the Hula Basin, Israel. *Israel Journal of Earth Science*, 22, 107–139.

Horowitz, A. (1977). The climate and the settlement in the Negev mountain during the last Quaternary. In *The Desert: Past, Present, Future*, ed. E. Sohar, pp. 52–59. Tel-Aviv: Reshafin Publishing (in Hebrew).

Horowitz, A. (1979). *The Quaternary of Israel*. New York: Academic Press.

Horowitz, A. (1980). Palynology – climate and distribution of settlements in Israel. *Kadmoniot*, 13, 51–52 (in Hebrew).

Horowitz, A. (1989). Continuous pollen diagrams for the last 3.5 M. Y. from Israel: vegetation, climate and correlation with the oxygen isotope record. *Palaeogeography, Palaeoclimatology and Palaeoecology*, 72, 63–78.

Hovan, S. A., Rea, D. K., Pisias, N. G. and Shackleton, N. J. (1989). A direct link between the China loess and marine d^{18}O records: aeolian flux to the north Pacific. *Nature*, 340, 296–298.

Huang, C. C., Zhou, J., Pang, J., Han, Y. and Hou, C. (2000). A regional aridity phase and its possible cultural impact during the Holocene megathermal in the Guanzhung Basin, China. *The Holocene*, 10, 135–142.

Huffman, T. N. (1996). Archaeological evidence for climatic change during the last 2000 years in Southern Africa. *Quaternary International*, 33, 55–60.

Hughes, M. K. and Graumlich, L. J. (1996). Multimillennial dendroclimatic studies from the western United States. In *NATO ASI Series*, vol. 141: *Climatic Variations and Forcing Mechanisms of the Last 2000 Years*, eds. P. D. Jones, R. S. Bradley and J. Jouzel, pp. 109–124. Berlin: Springer Verlag.

Hughes, P. D. M., Mauquoy, D., Barber, K. E. and Langdon (2000). Mire – development pathways and paleoclimatic records from a full Holocene peat archive at Walton Moss, Cumbria, England. *The Holocene*, 10, 465–479.

Hulme, M. (ed.) (1996). *Climate Change and Southern Africa: An Exploration of some Potential Impacts and Implications in the SADC Region*, Norwich, UK: Climatic Research Unit, University of East Anglia and WWF International, Switzerland, 104 pp.

Huntington, E. (1911). *Palestine and its Transformation*. Boston, MA: Houghton Mifflin, 443 pp.

Hyvärinen, H. and Albonen, P. (1994). Holocene lake-level changes in the Fennoscandian tree-line region, western Finnish Lapland: diatom and cladoceran evidence. *The Holocene*, 4, 251–258.

Isdale, P. J. and Kotwicki, V. S. (1987). Climate change scenarios for the Australian region. *South Australia Geographical Journal*, 87, 44–55.

Islebe, G. A., Hoogiemstra, H., Brenner, M., Curtis, J. H., Hodell, D. A. (1996). A Holocene vegetation history from lowland Guatemala. *The Holocene*, 6, 265–271.

Issar, A. (1968). Geology of the central coastal plain of Israel. *Israel Journal of Earth Science*, 17, 16–29.

Issar, A. (1979). Stratigraphy and paleoclimate of the Pleistocene of central and northern Israel. *Palaeogeography, Palaeoclimatology and Palaeoecology*, 29, 266–280.

Issar, A. (1990). *Water Shall Flow from the Rock*. Heidelberg: Springer Verlag, 213 pp.

Issar, A. S. (1995a). Climate change and the history of the Middle East. *American Scientist*, 83, 350–355.

Issar, A. S. (1995b). *Impacts of Climate Variations on Water Management and Related Socio-economic Systems*. [Technical Documents in Hydrology, IHP, IV Project H-2.1.] Paris: UNESCO, 95 pp.

Issar, A. (1997). Paleoclimate and paleokarst in South Africa. In *Proceedings of the 5th International Symposium and Field Seminar on Karst, Waters and Environmental Impacts* Antalya, Turkey, 10–12 September 1995, eds. G. Gunay and A. I. Johnson, pp. 265–267.

Issar, A. and Bruins, J. (1983). Special climatological conditions in the deserts of Sinai and Negev during the Late Pleistocene. *Palaeogeography, Palaeoclimatology and Palaeoecology*, 43, 63–72.

Issar, A. and Eckstein, Y. (1969). The lacustrine beds of Wadi Feiran, Sinai: their origin and significance. *Israel Journal of Earth Sciences*, 18, 21–27.

Issar, A. S. and Makover-Levin, D. (1995). Evidence for climatic changes during the time of the Bible in Israel. In *Mémoires de la Société Géologique de France*, No. 167: *Déserts Tropicaux et Changements Globaux*, pp. 67–71. Paris: Société Géologique de France.

Issar, A. and Tsoar, H. (1987). Who is to Blame for the Desertification of the Negev? In *Proceeding of the IAHS Symposium*, Vancouver, Canada, IAHS Publ. No. 168, pp. 577–583.

Issar, A. S. and Yakir, D. (1997). The Roman period's colder climate. *Biblical Archeologist*, 60, 2.

Issar, A. S., and Zohar, Z. (2003). *Climate, Water, Faith and Fate in the Near East*. Princeton University Press, in press.

Issar, A., Tsoar, H. and Levin, D. (1989). Climatic changes in Israel during historical times and their impact on hydrological pedological and socio-economic systems. In *Paleoclimatology and Paleometeorology*, eds. M. Leinen and M. Sarnthein, pp. 525–542, Dordrecht: Kluwer.

Issar, A. S., Govrin, Y., Geyh, M. E., Wakshal, E. and Wolf, M. (1992). Climate changes during the Upper Holocene in Israel. *Israel Journal of Earth Science*, 40, 219–223.

Iversen, J. (1973). *The Development of Denmark's Nature since the Last Glacial*. Geological Survey of Denmark V. Series No. 7–C. Copenhagen: Geological Survey of Denmark.

Jacobsen, T. (1957–58). *Salinity and Irrigation Agriculture in Antiquity, Diyala Basin Archaeological Project*. Bibliotheca Mesopotamica 14. Malibu: Udenda.

Jacobsen, T. (1960). The Waters of Ur, Iraq. *Science*, 22, 174–185.

Jacobsen, T. and Adams, R. M. (1958). Salt and silt in ancient Mesopotamian Agriculture. *Science* 128, 1251–1259.

Jäkel, D. (1987). Climatic fluctuations in the Central Sahara during the late Pleistocene and Holocene. In *The Climate of China and Global Climate, Beijing International Symposium on Climate*, 30 October to 3 November 1984 eds. Ye Dhuzheng, Fu Congbin, Chao Jiping and M. Moshino, pp. 124–137. Beijing: China Ocean Press and Springer Verlag.

Jauregui, E. (1997). Climate changes in Mexico during the historical and instrumented periods. *Quaternary International*, 43/44, 7–17.

Jelgersma, S., de Jong, J., Zagwijn, W. H. and van Regteren Altena, J. F. (1970). *The Coastal Dunes of Western Netherlands: Geology, Vegetational History and Archeology*, Mededelingen Rijks Geologische Dienst, Nieuwe Serie No. 21, pp. 94–152. The Hague: Geological Survey of the Netherlands.

Jennings, A. E. and Weiner, N. J. (1996). Environmental change in eastern Greenland during the last 1300 years: evidence from foraminifera and lithofacies in Nansen fjord, 68°N. *The Holocene*, 6, 179–191.

Ji, J., Petit-Maire, N. and Yan, Z. (1993). The last 1000 years: climatic change in arid Asia and Africa. *Global and Planetary Change*, 7, 203–210.

Jingtai, W. and Kaqin, J. (1989). Geomorphology, Quaternary sedimentary and lake level fluctuations in the Chaiwopu Basin, Xingjiang, China. In *Water Resources and Environment in Chaiwopu–Dabancheng Region*, eds. Shi Yafeng *et al.*, pp. 11–22. Beijing: Congress Science Press.

Jinjun Je, Zhongwei Yan and Yongquiang Liu (1990). *Climatic Change in Chinese Deserts*. Beijing: Institute of Atmospheric Physics, Academica Sinica.

Johnson, D. L. (1969). *The Nature of Nomadism: A Comparative Study of Pastoral Migrations in Southwestern Asia and Northern Africa*. [Research Paper No. 118.] Chicago, IL: Chicago University Press.

Johnson, J. B., Gifford, H. M., Fogel, L. M. and Beaumont, B. P. (1997). The determination of late Quaternary paleoenvironments at Equus Cave, South Africa, using stable isotopes and amino acid racemization in ostrich eggshell. *Palaeogeography, Palaeoclimatology and Palaeoecology*, 136, 121–137.

Jus, M. (1982). Swiss Midland-lakes and climatic changes. In *Climatic Changes in Later Prehistory*, ed. A. F. Harding, pp. 44–51. Edinburgh: Edinburgh University Press.

Kalicki, T. (1991). The evolution of the Vistula River valley between Cracow and Niepolomice in Late Vistulian and Holocene Times. In *Geographical Studies*, Special Issue 6: *Evolution of the Vistula Valley During the Last 15 000 Years*. pp. 11–38. Warsaw: Polish Academy of Science.

Karlén, W. (1991). Glaciers fluctuations in Scandinavia, in temperate paleohydrology. In *Fluvial Processes in the Temperate Zone during the last 15 000 years*, eds. L. Starkel, K. J. Gregory, and J. B. Thornes, pp. 395–412. Chichester, UK: John Wiley & Sons.

Karlén, W. and Kuylenstierna, J. (1996). On solar forcing of Holocene climate: evidence from Scandinavia. *The Holocene*, 6, 359–364.

Kaufman, A., Wassetburg, G. J., Porcelli, D., Bar-Matthews, M., Ayalon, A. and Halicz, L. (1998). U-Th isotope systematics from the Soreq cave, Israel and climatic correlations. *Earth Planet Science Letters*, 156, 141–155.

Kay, P. A. (1979). Multivariate statistical estimates of Holocene vegetation and climate change, forest–tundra transition zone, NWT, Canada. *Quaternary Research*, 11, 125–140.

Kay, P. A. and Johnson, D. L. (1981). Estimation of Tigris–Euphrates streamflow from regional paleoenvironmental proxy data. *Climatic Change*, 3, 251–263.

Kedar, B. Z. (1985). The Arab conquest and agriculture, a seventh century apocalypse, satellite imagery and palynology. *Asian and African Studies*, 19, 1–15.

Kellogg, T. B. (1984). Late glacial Holocene high frequency climatic changes in deep sea cores from the Denmark Strait. In *Climatic Changes on a Yearly to Millennial Basis*, eds. N. A. Mörner and W. Karlen, pp. 123–134. Dordrecht: Reidel.

Kenyon, K. M. (1957). *Digging up Jericho*. London: Benn. 272 pp.

Kenyon, K. M. (1979). *Archaeology in the Holy Land*. New York: Norton.

Kenyon, K. M., Bottero, J. and Posener, G. (1971). Syria and Palestine c. 2160–1780 B.C. In *Cambridge Ancient History*, 3rd revised edn, vol. I (2), eds. I. E. S. Edwards, C. J. Gadd and N. G. L Hammond, pp. 532–594. Cambridge: Cambridge University Press.

Kezau, C. and Bowler, J. M. (1985). Preliminary research on the sedimentary characteristics of Chaeran salt lake in Qaidan Basin and the paleoclimatic evolution. *Scientia Sinica* (Series B). 5, 463–473.

Kim, G. S. and Choi, I. S. (1987). A preliminary study on long term variations of unusual climate phenomena during the past 1000 years in Korea. In *The Climate of China and Global Climate, Beijing International Symposium on Climate*, 30 October to 3 November 1984, eds. Ye Dhuzheng, Fu Congbin, Chao Jiping and M. Moshino, pp. 30–37. Bejing: China Ocean Press and Springer Verlag.

Klein, C. (1982). Morphological evidence of lake level changes on the western shores of the Dead Sea. *Israel Journal of Earth Sciences*, 31, 67–94.

Klein, J., Lerman, J. C., Damon, P. E. and Ralph, E. K. (1982). Calibration of radiocarbon dates. *Radiocarbon*, 24, 103–150.

Klein, R., Loya, Y., Gvirtzman, G., Isdale, P. J. and Susie, M. (1990). Seasonal rainfall in the Sinai Desert during the Late Quaternary inferred from fluorescent bands in fossil corals. *Nature*, 345, 145–147.

Knox, J. C. (1993). Large increase in flood magnitude in response to modest changes in climate. *Nature*, 361, 430–432.

Kochavi, M. (1969). The Middle Bronze Age I (The Intermediate Bronze Age) in Eretz Israel. *Qadmoniot*, 2, 38–44 (in Hebrew).

Koizumi, T. and Aoyagi, S. (1993). Debris supply periods estimated from weathering rind thickness in rubble on the west-facing slope of Mt. Yakushidake, the Northern Japan Alps. *Geographical Review of Japan*, 66A, 269–286.

Koopmann, B. (1981). Sedimentation von saharastaub im subtropischen Atlantik während der letzten 25,000 Jabre. *"Meteor" Forschung Ergebnisse*, C 35, 23–59.

Kukla, G. and An Zhisheng (1989). Loess stratigraphy in Central China. *Palaeogeography, Palaeoclimatology and Palaeoecology*, 72, 203–225.

Kukla, G. J., Angell, J. K., Korshover, J. *et al.* (1977). New data on climatic trends. *Nature*, 270, 573–577.

Kutzbach, J. E. (1983). Modeling of Holocene climates. In *The Holocene: Late Quaternary Environments of the United States*, vol. 2, ed. H. E. Wright, Jr, pp. 271–277. Minneapolis: University of Minnesota Press.

Kutzbach, J. E. and Guetter, P. J. (1986). The influence of changing orbital parameters and surface boundary conditions on climate simulation for the past 18 000 years. *Journal of Atmospheric Sciences*, 43, 1726–1759.

Kutzbach, J. E. and Street-Perrott, F. A. (1985). Milankovitch forcing of fluctuation in the level of tropical lakes from 18 to 0 K yr B.P. *Nature*, 317, 130–134.

LaMarche, V. C. (1974). Paleoclimatic inferences from long tree-ring records. *Science*, 183, 1043–1048.

Lamb, H. H. (1968). Volcanic dust, melting of ice caps and sea levels – discussion. Bloch, M. R. – A reply. *Palaeogeography, Palaeoclimatology and Palaeoecology*, 4, 219–226.

Lamb, H. H. (1977). *Climate: Present Past and Future*, vol. II. London: Methuen, 835 pp. [Paperback edition (1985). Princeton University Press.]

Lamb, H. H. (1982). *Climate, History and the Modern World*. London: Methuen, 387 pp.

Lamb, H. H. (1984a). Climate and history in northern Europe and elsewhere. In *Climatic Changes on a Yearly to Millennial Basis*, eds. N. A. Mörner and W. Karlen, pp. 225–240. Dordrecht: Reidel.

Lamb, H. H. (1984b). Some studies of the Little Ice Age of recent centuries and its great storms. In *Climatic Changes on a Yearly to Millennial Basis*, eds. N. A. Mörner and W. Karlen, pp. 309–329. Dordrecht: Reidel.

Lancaster, I. N. (1979). Evidence for a widespread late Pleistocene humid period in the Kalahari. *Nature*, 279, 145.

Lapp, P. (1970). Palestine in the Early Bronze Age. In *Near Eastern Archaeology in the 20th Century – Essays in Honor of Nelson Glueck*, ed. J. A. Sanders, pp. 101–131. New York: Doubleday.

Lario, J., Zazo, C., Dabrio, C. J. *et al.* (1995). Record of recent Holocene sediment input on spit-bars and deltas in South Spain. [In *Holocene cyclic pulses and sedimentation*, ed. B. Core.] *Journal of Coastal Research, Special Issue*, 17, 201–205.

Larsen, C. E. and Evans, G. (1978). The Holocene geological history of the Tigris Euphrates–Karun Delta. In *The Environmental History of the Near and Middle East Since the Ice Age*, ed. W. C. Brice, pp. 227–245. London: Academic Press.

Lauer, W. and Frankenberg, P. (1980). Modelling of climate and plant cover in the Sahara for 5500 B.P. and 18,000 B.P. *Palaeoecology of Africa*, 12, 307–314.

Leavitt, S. W. (1994). Major wet interval in White Mountains Medieval warm period evidenced in d^{13}C of Bristlecone pine tree rings. In *The Medieval Warm Period*, eds. M. K. Hughes and H. F. Diaz, pp. 299–307. Dordrecht: Kluwer Academic.

Leemann, A. and Niessen, F. (1994). Holocene glacial activity and climatic variations in the Swiss Alps: reconstructing a continuous record from proglacial lake sediments. *The Holocene* 4, 259–268.

Leguy, C., Rindsberger, M., Zangvil, A., Issar, A. and Gat, J. (1983). The relation between the oxygen 18 and deuterium contents of rain water in the Negev Desert and air-mass trajectories. *Isotope Geoscience*, 1, 205–218.

Lemcke, G. and Sturm, M. (1997). d^{18}O and trace element as proxy for the reconstruction of climate changes at Lake Van (Turkey): preliminary results. In *NATO ASI Series*, vol. 149: *Third Millennium BC Climate Change and Old World Collapse*, eds. N. Dalfes, G. Kukla and H. Weiss, pp. 654–678. Berlin: Springer Verlag.

Leroi-Gourhan, A. (1974). Etudes palynologiques des derniers 11,000 ans en Syrie semi-desertique. *Paléorient*, 2, 443–451.

Leroi-Gourhan, A. (1980). Les analyses polliniques au Moyen-Orient. *Paléorient*, 6, 79–91.

Leroi-Gourhan, A. (1981). Diagrammes polliniques de sites archéologiques au Moyen Orient. In *Beihefte zum Tübinger Atlas des Vorderen Orients*, vol. 8, eds. W. Frey, H.P. Uerpmann and A. Reihe, pp. 121–133. Tubingen: Beiträge zur Umweltgeschichte des vorderen Orients.

Leroi-Gourhan, A. and Darmon, F. (1987). Analyses palynologiques de sites archéologiques du Pléistocene final dans la Vallée du Jourdain. *Israel Journal of Earth Sciences*, 36, 65–72.

Le Roy Laudurie, E. (1971). *Times of Feast, Times of Famine. A History of Climate Since the Year 1000*. New York: Doubleday, 426 pp.

Lev-Yadun, S., Liphschitz, N. and Waisel, Y. (1987). Annual rings in trees as an index to climate changes intensity in our region in the past. *Rotem*, 22: 6–17 (in Hebrew). 113 (English abstract).

Levy, T. E. (1986). The Chalcolithic period. *Biblical Archaeologist*, June, 83–108.

Levy, T. E. (1998). Cult, metallurgy and rank societies – Chalcolithic Period (*c.* 4500–4350 B.C.E.). In *The Archaeology of Society in the Holy Land*, ed. T. E. Levy, pp. 226–244. London: Leicester University Press.

Lewis, N. (1948). New light on the Negev in ancient times. *Palestine Exploration Quaternary*, 80, 102–117.

Lézine, A.-M. (1998). Pollen records of past climate changes in West Africa since the Last Glacial maximum. In *Water, Environment and Society in Times of Climate Change*, eds. A. Issar and N. Brown, pp. 295–319. Dordrecht: Kluwer Academic.

Lézine, A.-M. and Casanova, J. (1989). Pollen and hydrological evidence for the interpretation of past climates in tropical west Africa during the Holocene. *Quaternary Science Reviews*, 8, 45–55.

Lézine, A.-M., Casanova, J. and Hillaire-Marcel, C. (1990). Across an early Holocene humid phase in western Sahara: pollen and isotope stratigraphy. *Geology*, 18, 264–267.

Li Jijun (1990). The patterns of environmental changes since Late Pleistocene in northwestern China. The impact of climatic change on the Chinese migrations in historical times. In *Regional Conference on Asian Pacific Countries of I.G.U., Global Change and Environmental Evolution in China*, Section III, eds. Liu Chuang, Zhao Songqiao, Zhang Peiyuan and Shi Peijun, pp. 19–25. Hohot, P. R. China: Editorial Board of Arid Land Resources.

Lin Zhenyao (1990). Climatic Change in Tibet during the last 200 years. In *Climatic Changes and Their Impacts, Beijing International Symposium on Climatic Change*, 9–12 August, p. A-8. Beijing: Natural Science Foundation of China (CNSF).

Lin Zhenyao and Wu Xianding (1987). Some features of climatic fluctuations over the Qinghai-Xizang Plateau of Tibet. In *The Climate of China and Global Climate, Beijing International Symposium on Climate*, 30 October to 3 November 1984, eds. Ye Dhuzheng, Fu Congbin, Chao Jiping and M. Moshino, pp. 116–123. Beijing: China Ocean Press and Springer Verlag.

Liphschitz, N. and Waisel, Y. (1974). The effects of human activity on composition of the natural vegetation during historic periods. *Le-Yaaran*, 24, 9–15 (in Hebrew). 27–30 (English).

Liphschitz, N., Lev-Yadun, S. and Waisel, Y. (1979a). Dendrochronological investigations in the Mediterranean Basin – *Pinus nigra* of south Anatolia (Turkey). *La-Yaaran*, 29, 1–10. [The Journal of the Israel Forestry Association.]

Liphschitz, N., Waisel, Y. and Lev-Yadun, S. (1979b). Dendrochronological investigations in Iran. *Tree Ring Bulletin*, 39, 39–45.

Liphschitz, N., Lev-Yadun, S. and Waisel, Y. (1981). Dendroarchaeological investigations in Israel (Masada). *Israel Exploration Journal*, 31, 230–234.

Li Pingri and Fang Guoxiang (1991). Quaternary deposits and environmental evolution in the Guangzhou Plain and Zhuziang Delta. In *Quaternary Geology and Environment in China, Series of the XIII INQUA Congress*, ed. Liu Tungsheng, pp. 390–395. Beijing: Congress Science Press.

Li Shuan-Ke (1990). Fluctuations of closed lake level, and variations of Paleo climatology in North Tibetan Plateau China. In *Regional Conference on Asian Pacific Countries of I.G.U., Global Change and Environmental Evolution in China*, Section III, eds. Liu Chuang, Zhao Songqiao, Zhang Peiyuan and Shi Peijun, pp. 53–61. Hohot, P. R. China: Editorial Board of Arid Land Resources.

Liu Changming and Fu Goubin (1996). The impact of climatic warming on hydrological regimes in China: an overview. In *The Geojournal Library*, vol. 38: *Regional Hydrological Response to Climate Change*, eds. J. A. A. Jones, Changming Liu, Ming-Ko Woo and Hsiang-Te Kung, pp. 133–153. Dordrecht: Kluwer.

Liu Weilun and Wang Yunzhang (1990). Some causes of drought in the Yellow River valley. In *Climatic Changes and Their Impacts, Beijing International Symposium on Climatic Change*, 9–12 August, p. 21. Beijing: Natural Science Foundation of China (CNSF).

Liver, Y. (1986). The Diaspora in Babylon, the return to Zion and the Persian reign. In *The History of the Land of Israel*, vol. 1, ed. Y. Ripel, pp. 207–222. Tel Aviv: Israeli Ministry Publishing (in Hebrew).

Lloyd, A. H. and Graumlich, L. J. (1997). Holocene dynamics of treeline forests in the Sierra Nevada. *Ecology*, 78, 1199–1210.

Lopez-Vera, F. (ed.) (1986). *Proceedings of the Symposium of Climatic Fluctuations during the Quaternary in the Western Mediterranean Regions*. Madrid: Universidad Autonoma de Madrid, 563 pp.

Lough, J. M., Fritts, H. C. and Wu Xiangchi (1987). Relationship between the climates of China and North America over the past four centuries, a comparison of proxy data. In *The Climate of China and Global Climate, Beijing International Symposium on climate*, 30 October to 3 November 1984, eds. Ye Dhuzheng, Fu Congbin, Chao Jiping and M. Moshino, pp. 89–105. Bejing: China Ocean Press and Springer Verlag.

Louwe Kooijmans, L. P. (1974). The Rhine/Meuse delta; four studies on its prehistoric occupation and Holocene geology. In *Anale Prehistorica Leidensia*, 7. Leiden: University of Leiden.

Louwe Kooijmans, L. P. (1980). Archaeology and the coastal change in the Netherlands. In *Archeology and Coastal Change*, ed. F. H. Thompson. London: Society of Antiquaries.

Lozano-Garcia, M. and Xelhuantzi-Lopez, M. S. (1997). Some problems in the Late Quaternary pollen records of central Mexico: basins of Mexico and Zacapu. *Quaternary International*, 43/44, 117–123.

Luz, B. (1979). Paleo-oceanography of the post-glacial eastern Mediterranean. *Nature*, 278, 847–848.

Luz, B. (1991). Post-glacial Paleo-oceanography of the Eastern Mediterranean: *International Workshop on Regional Implications of Future Climate Change*. 28 April to 2 May 1991. Ministry of Environment. Jerusalem: Israel.

Luz, B. and Perelis-Grossowicz, L. (1980). Oxygen isotopes, biostratifigraphy and recent rates of sedimentation in the Eastern Mediterranean off Israel. *Israel Journal of Earth Sciences*, 29, 140–146.

Maejima, I. (1980). Seasonal and regional aspects of Japan's weather and climate. In *Geography of Japan*, ed. The Association of Japanese Geographers. Tokyo: Teikoku-Shoin.

Magaritz, M. and Goodfriend, G. A. (1987). Movement of the desert boundary in the Levant from latest Pleistocene to early Holocene. In *Abrupt Climatic Change – Evidence and Implications*, eds. W. H. Berger and L. D. Labeyrie, pp. 173–183. Dordrecht: Reidel.

Magny, M. (1992). Holocene lake-level fluctuations in Jura and the northern subalpine ranges, France: regional pattern and climatic implications. *Boreas*, 21, 319–334.

Maley, J. (1977a). Palaeoclimates of Central Sahara during the early Holocene. *Nature*, 269, 573–577.

Maley, J. (1977b). Analyses polliniques et paléoclimatologie des douze derniers millénaires du bassin du Tchad (Afrique centrale). *Supplement au Bulletin AFEQ*, 1977–1, No. 50, 187–197.

Maley, J. (2000). Last Glacial maximum lacustrine and fluviatile formations in Tibesti and other Saharan mountains, and large scale climatic teleconnections linked to the activity of the subtropical jet stream. *Global and Planetary Change*, 26, 121–136.

Maley, J. and Livingstone, D. A. (1983). Extension d'un élément montagnard dans le sud du Ghana (Afrique de l'Ouest) au Pléistocène supérieur et à l'Holocène inférieur: Premierer données polliniques. *Comptes Rendus des Séances de l'Académie des Sciences, Paris*, 296, Serie II, 1287–1292.

Mamedov, A.V. (1997). The Late Pleistocene–Holocene history of the Caspian Sea. *Quaternary International*, 41/42, 161–166.

Mangerud, J. (1987). The Alleröd/Younger Drias boundary. In *Abrupt Climatic Change*, eds. W. H. Berger and L. D. Labeyrie, pp. 163–171. Dordrecht: Reidel.

Manzanilla, L. (1997). The impact of climatic change on past civilizations. A revisionist agenda for further investigations. *Quaternary International*, 43/44, 153–159.

Mayerson, P. (1983). The city of Elusa in the literary sources of the fourth-sixth centuries. *Israel Exploration Journal*, 33, 247–253.

Mazar, B. (1967). The Middle Bronze Age in the Land of Israel. *Eretz Israel*, 8, 216–230 (in Hebrew).

Mazar, B. (1968). The Middle Bronze Age in Palestine. *Israel Exploration Journal*, 18, 65–97.

Mazar, B. (1986). The Land of Israel in the Middle Bronze Age. In *The History of the Land of Israel*, vol. 1, ed. Y. Ripel, pp. 73–86, Tel Aviv: Israeli Ministry Publishing (in Hebrew).

Mazar, A. (1990). The Iron Age I. In *The Archaeology of Ancient Israel*, ed. A. Ben-Tor, pp. 258–301. Tel Aviv: Open University of Israel; New Haven, CT: Yale University Press.

McKenzie, J. A. and Eberli, G. P. (1987). Indications for abrupt Holocene climatic change: Late Holocene oxygen isotope stratigraphy of the Great Salt Lake, Utah. In *Abrupt Climate Change*, ed. W. H. Berger and L. D. Labeyrie, pp. 127–136. Dordrech: Reidel.

Mellaart, J. (1966). *The Chalcolithic and Early Bronze Ages in the Near East and Anatolia*, pp. 1–57. Beirut: Khayats.

Menéndez-Amor, J. and Florschütz, F. (1961). Resultados del analisis polinico de una serie de turba recogida en la ereta de Pedregal (Navarres, Valencia). *Archivo de Prehistoria Levantina*, 9, 97–99.

Meshel, Z. (1977). The Negev during the Persian period. *Katedra*, 4, 43–50 (in Hebrew).

Metcalfe, S. E. (1997). Paleolimnological records of climate change in Mexico – frustrating past, promising future? *Quaternary International*, 43/44, 11–116.

Michel, P. (1980). The SW Sahara margin: sediments and climatic changes during the recent Quaternary. *Palaeoecology of Africa*, 12, 297–306.

Mikami, T. (1987). Climate of Japan during 1781–90 in comparison with that of China. In *The Climate of China and Global Climate, Beijing International Symposium on Climate*, 30 October to 3 November 1984, eds. Ye Dhuzheng, Fu Congbin, Chao Jiping and M. Moshino, pp. 64–75. Beijing: China Ocean Press and Springer Verlag.

Miller, R. (1980). Water use in Syria and Palestine from the Neolithic to the Bronze Age. *World Archaeology*, 11, 331–341.

Miyoshi, N. and Yano, N. (1986). Late Pleistocene and Holocene vegetational history of the Ohnuma Moor in the Chugoku Mountains, western Japan. *Review of Paleobotany and Palynology*, 46, 355–376.

Mörner, N. A. (1978–79). The northwest European "sea-level laboratory" and Regional Holocene eustasy. *Palaeogeography, Palaeoclimatology and Palaeoecology*, 29, 281–300.

Mörner, N. A. (1984). Climatic changes on a yearly to millennial basis – an introduction. In *Climatic Changes on a Yearly to Millennial Basis*, eds. N. A. Mörner and W. Karten, pp. 1–13. Dordrecht: Reidel.

Mörner, N. A. and Wallin, B. (1977). A 10,000- year temperature record from Gotland Sweden, *Palaeogeography, Palaeoclimatology and Palaeoecology*, 21, 113–138.

Muzzolini, A. (1986). Archeologie Africaine et Sciences de la Nature applique a l'Archeologie. In *First International CNRS Symposium*, Bordeaux 1983, pp. 53–69. Paris: CNRS.

Neeman, N. (1982). The land of Israel during the Canaanite period: the Middle Bronze Age and the Late Bronze Age (ca. 2000–1200 BC). In *The History of the Land of Israel in The Ancient Periods*, ed. Y. Efal, pp. 131–255. Jerusalem: Keter, Yad Yzhak Ben-Zvi (in Hebrew).

Neev, D. and Emery, K. O. (1967). *Geological Survey of Israel Bulletin* 41: *The Dead Sea, depositional processes and environments of deposition.* Jerusalem: Geological Survey of Israel.

Neev, D. and Hall, J. K. (1977). Climatic fluctuations during the Holocene as reflected by the Dead Sea levels. In *International Conference on Terminal Lakes*, Weber State College, Ogden, Utah, 2–5 May (preprint).

Negev, A. (1965). The history of the Nabatian kingdom. *Mada*, 9, 1–9 (in Hebrew).

Negev, A. (1977). The beginning of the Nabatian kingdom. In *Memorial Papers to Amnon: Between Hermon and Sinai*, pp. 143–162, Jerusalem: Ronald Publishing (in Hebrew).

Negev, A. (1979). The Nabatian in the Negev. In *The Land of the Negev – A Man and a Desert*, vol. 1, eds. A. Shmueli and Y. Grados, pp. 226–269. Tel Aviv: Israeli Defense Ministry Publishing (in Hebrew).

Nesje, A. and Dahl, S. O. (1991). Holocene glacier variations of Blaisen, Hardangerjokulem, central southern Norway. *Quaternary Research*, 35, 25–40.

Neuman, J. (1985). Climatic change as a topic in the classical Greek and Roman Liturature. *Climatic Change*, 7, 441–454.

Neuman, J. and Parpola, S. (1987). Climatic change and the eleventh–tenth century eclipse of Assyria and Babylonia. *Journal of Near Eastern Studies*, 46, 161–182.

Neuman, J. and Sigrit, R. M. (1978). Harvest dates in ancient Mesopotamia as possible indicators of climatic variations. *Climatic Change*, 1, 239–252.

Newell, R. E. and Hsiung, J. (1987). Factors controlling free air and ocean temperature of the last 30 years and extrapolation to the past. In *Abrupt Climatic Change*, eds. W. H. Berger and L. D. Labeyrie, pp. 67–87. Dordrecht: Reidel.

Nicholson, S. E. (1980). Saharan climates in historic times. In *Sahara and the Nile*, eds. M. A. J. Wiliams and H. J. Faure, pp. 173–200. Rotterdam: Balkema.

Nicholson, S. E. and Flohn, H. (1980). African evironmental and climatic changes and the general circulation in the Late Pleistocene and Holocene. *Climatic Change*, 2, 313–348.

Nigam, S., Barlow, M. and Berbery, E. (1999). Analysis links pacific decadal variability to drought and streamflow in United States. *EOS Transactions American Geophysical Union*, 80, 621–625.

Nir, Y. (1997). Middle and late Holocene sea-levels along the Israel Mediterranean coast – evidence from ancient wells. *Journal of Quaternary Science*, 12, 141–151.

Nir, Y. and Eldar, I. (1987). Ancient wells and their geoarchaeological significance in detecting tectonics of the Israel Mediterranean coast line region. *Geology*, 15, 3–6.

Oeschger, H., Beer, J., Siegenthaler, U., Stauffer, B., Dansgaard, W. and Langway, C. C. (1984). Late glacial climate history from ice cores. In *Climate Processes and Climate Sensitivity*, eds. J. E. Hansen and T. Takahashi, pp. 299–305. Washington, DC: American Geophysical Union.

Oguchi, T. (1988). Landform development during the Last Glacial and the Post-Glacial ages in the Matsumoto Basin and its surrounding mountains, central Japan. *Quaternary Research* (Tokyo), 27, 101–124.

Oguchi, T. (1997). Channel incision and sediment production in Japanese mountains, in relation to past and future climate change. In *Proceedings of the Conference on Management of Landscapes Disturbed by Channel Incision*, Oxford, Mississippi, pp. 867–872.

O'Hara, S. L. and Metcalfe, S. E. (1997). The climate of Mexico since the Aztec Period. *Quaternary International*, 43/44, 25–31.

Ono, Y. and Hirakawa K. (1975). Glacial and periglacial morphogenetic environments around the Hidaka Range in the Würm Glacial age. *Geographic Reviews of Japan*, 48, 1–26.

Otterman, J. (1974). Baring high-albedo soils by overgrazing: a hypothesized desertification mechanism. *Science*, 86, 531–533.

Owen, R. B., Barthelme, J. W., Renaut, R. W. and Vincens, A. (1982). Paleolimnology and archaeology of Holocene deposits north-east of Lake Turkana, Kenya. *Nature*, 298, 523–529.

Pachur, H. J. and Braun, G. (1980). The paleoclimate of the Central Sahara, Libya and the Libyan desert. *Paleoecology of Africa*, 12, 351–363

Paepe, L. (1984). Landscape changes in Greece as a result of changing climate during the Quaternary in Desertification in Europe. In *Proceedings of the International Symposium in the EEC Programme on Climatology*, Mytilene, Greece, 15–18 April, eds. R. Fantechi and N. S. Margaris, pp. 2–25. Dordrecht: Reidel.

Palmer, T. N. (1986). Gulf Stream variability and European climate. *Meteorological Magazine*, 115, 291–297.

Pan Tiefu (1990). Climate becoming warmer and drier in the Jilin Province of China. In *Climatic Changes and Their Impacts, Beijing International Symposium on Clime*, 9–12 August, p. c-7. Beijing: Natural Science Foundation of China (CNSF).

Parra, I. (1994). Quantification des precipitations a partir des spectres polliniques actuels et fossiles: du tardiglaciaire a l'Holocene Superieur de la Cote Mediterranneene Espagnole. PhD Thesis, L'Ecole Pratique de Hautes Etude, Ministère de L'Enseignment Superieur et de la Rechecte, Montpellier, France.

Partridge, T. C., Avery, D. M., Botha, G. A. *et al.* (1990). Pleistocene and Holocene climatic change in Southern Africa. *South Africa Journal of Science*, 86, 302–306.

Partridge, T. C., Kerr, S. J., Metcalfe, S. E., Scott, L., Talma, A. S. and Vogel, J. C. (1993). The pretoria saltpan: a 200 000 years Southern African lacustrine sequence. *Palaeogeography, Palaeoclimatology and Palaeoecology* 10, 317–337.

Perez-Obiol, R. and Julia, R. (1994). Climatic change on the Iberian peninsula recorded in a 30,000-yr pollen record from Lake Banyoles. *Quaternary Research*, 41, 91–98.

Perrot, J. (1968). La prehistoire Palestinienne. In *Supplement au Dictionnaire de la Bible* VIII, pp. 416–438. Paris: Edition Letouzey et Ane.

Petersen, K. L. (1994). A warm and wet little climatic optimum and a cold and dry Little Ice Age in the Southern Rocky Mountains. In *The Medieval Warm Period*, eds. M. K. Hughes and H. F. Diaz, pp. 243–249. Dordrecht: Kluwer Academic.

Petit-Maire, N. (1980a). Holocene biogeographical variations along the NW African coast (28°–19°N): paleoclimatic implications. *Palaeoecology of Africa*, 12, 365–377.

Petit-Maire, N. (1980b). Pleistocene lakes in the Shati area, Fezzan (27° 30' N). *Palaeoecology of Africa*, 12, 289–295.

Petit-Maire, N. (1987). Climatic evolution of the Sahara in northern Mali during the Recent Quaternary. In *The Climate of China and Global Climate, Beijing International Symposium on Climate*, 30 October to 3 November 1984, eds. Ye Dhuzheng, Fu Congbin, Chao Jiping and M. Moshino, pp. 135–137. Beijing: China Ocean Press and Springer Verlag.

Picard, L. (1943). *Structure and Evolution of Palestine.* Jerusalem: Hebrew University, 187 pp.

Pirazzoli, P. A. and Delibrias, G. (1983). Late Holocene and recent sea level changes and crustal movement in Kume Island, the Ryukus Japan. *Bulletin of the Department of Geography, University of Tokyo*, 15, 63–76.

Pisias, N. G. (1978). Paleoceanography of the Santa Barbara Basin during the last 8000 years. *Quaternary Research*, 10, 366–384.

Pisias, N. G. (1979). Model for paleoceanographic reconstruction of the California current during the last 8000 years. *Quaternary Research*, 11, 373–386.

Pollock, S. (1999). *Ancient Mesopotamia: The Eden that Never Was.* Cambridge: Cambridge University Press, 259 pp.

Pons, A. and Reille, M. (1986). Nouvelles recherches du pollen a Padul (Granada). La fin du dernier glaciation a L'Holocene. In *Proceedings of the Symposium of Climatic Fluctuations during the Quaternary in the Western Mediterranean Regions*, ed. F. Lopez-Vera, pp. 405–422. Madrid: Universidad Autonoma de Madrid.

Pons, A. and Reille, M. (1988). The Holocene and Upper Pleistocene pollen record from Padul (Grananda Spain): a new study. *Palaeogeography, Palaeoclimatology and Palaeoecology*, 66, 243–263.

Pons, A., de Beaulieu, J. L., Guiot, J. and Reille, M. (1987). The Younger Dryas in southwestern Europe: an abrupt climatic change as evidenced from pollen records. In *Abrupt Climatic Change*, eds. W. H. Berger and L. D. Labeyrie, pp. 195–208. Dordrecht: Reidel.

Porter, S. C. (1981). Glaciological evidence of Holocene climatic change. In *Climate and History*, eds. T. M. L. Wigley, M. J. Ingram and G. Farmer, pp. 82–110. Cambridge: Cambridge University Press.

Porter, S. C. (1986). Pattern and forcing of Northern Hemisphere glacier variations during the last millennium. *Quaternary Research*. 26, 27–48.

Ponyxi, Z., Baozhen, Z. and Wenbo, Y. (1988). The evolution of the water body environment in Qinghai Lake, since the Postglacial Age. *Acta Sedimentologica Sinica*, 16, 1–13.

Prell, W. L. and Kutzbach, J. E. (1987). Monsoon variability over the past 150,000 years. *Journal of Geophysical Research*, 92(D7), 8411–8425.

Pye, K. (1987). *Aeolian Dust and Dust Deposits*. London: Academic Press.

Raban, A. (1991). Evidence of sea–land vertical changes from archaeological sites. In *Proceedings of the Israeli Geological Society Conference*, Acare, April, Guide 4, pp. 131–811 (in Hebrew).

Raban, A. and Galili, E. (1985). Recent maritime archaeological research in Israel: a preliminary report. *International Journal of Nautical Archaeology and Underwater Exploration*, 14, 321–356.

Reed, J. M., Stevenson, A. C. and Juggins, S. (2001). A multi-proxy record of Holocene climatic change in southwestern Spain: the laguna de Medina, Cádiz. *The Holocene*, 11, 707–719.

Richard, S. (1980). Toward a consensus of opinion on the end of the Early Bronze Age in Palestine–Transjordan. *Bulletin of American Schools of Oriental Research*, 237, 5–34.

Richard, S. (1987). the Early Bronze Age: the rise and collapse of urbanism. *Biblical Archaeologist*, 50, 22–43.

Riquelem Cantal, R. M. (1994). Consideraciones a traves de la fauna sobre la economia y el medio ambiente del yacimiento de Acinipo, Ronda (Malaga) durante la primera mitad del II milenio A. C. Thesis de Liecent, University of Granada.

Ritchie, J. C. and Haynes, C. V. (1987). Holocene vegetation zonation in the eastern Sahara. *Nature*, 330, 645–647.

Ritter-Kaplan, H. (1984). The impact of drought on third millennium culture on the basis of excavations in the Tel Aviv grounds. In *The Land of Israel, Braver Book*, 17, pp. 333–338. Jerusalem: Society for Investigation of Eretz Israel and its Antiquities (in Hebrew).

Roberts, N. (1989). *The Holocene – An Environmental History*. Oxford: Basil Blackwell, 227 pp.

Rodó, X., Baert, E. and Comin, F. A. (1997). Variations in seasonal rainfall in Southern Europe during the present century: relationships with the North Atlantic oscillation and the El Niño–Southern oscillation. *Climate Dynamics*, 13, 245–284.

Roeleveld, W. (1974). The Groningen Coastal Area: A Study in Holocene Geology and Low-land Physical Geography. *Berichten van de Riksedienst voor Oudheidkundig Bodemonderzek*, vol. 24, Amsterdam: Government Publishing Agency.

Rognon, P. (1976). Constructions alluviales holocenes et oscillations climatiques dans Sahara meridional. *Bulletin of the French Geographical Association*, 433, 77–84.

Rognon, P. (1987a). Aridification and abrupt climatic events on the Saharan northern and southern margins, 20,000 Y B.P. to present. In *Abrupt Climatic Change*, eds. W. H. Berger and L. D. Labeyrie, pp. 209–220. Dordrecht: Reidel.

Rognon, P. (1987b). Late Quaternary climatic reconstruction for the Maghreb (North Africa). *Palaeogeography, Palaeoclimatology and Palaeoecology* 58, 11–34.

Rohrlich, V. and Goldsmith, V. (1984). Sediment transport along the southeast Mediterranean: a geological perspective. *Geo-Marine Letters*, 4, 99–103

Rosen, A. M. (1986). Environment and culture at Tel Lachish, Israel. *Bulletin of the American Schools of Oriental Research*, 263, 55–60.

Rosen, A. M. (1997). Environmental change and human adaptational failure at the end of the Early Bronze Age in the Southern Levant. In *NATO ASI Series*, vol. 149, *Third Millennium BC Climate Change and Old World Collapse*, eds. N. Dalfes, G. Kukla and H. Weiss, pp. 25–38. Berlin: Springer.

Rossignol-Strick, M., Nesteroff, W., Olive P. and Vergnaud-Grazzini, C. (1982). After the deluge: Mediterranean stagnation and sapropel formation. *Nature*, 295, 105–110.

Ruddiman, W. G. and Duplessy, J. C. (1985). Conference on the last deglaciation: timing and mechanism. *Quaternary Research*, 23, 1–17.

Ruddiman, W. G. and McIntyre, A. (1981). The North Atlantic Ocean during the last deglaciation. *Palaeogeography, Palaeoclimatology and Palaeoecology*, 35, 145–214.

Ruiz Bustos, A. (1995). Quantification of the climatic conditions of Quaternary sites by means of mammals. In *IX Reunion Nacional Sobre Cuaternario: Reconstruccion de Paleoambientes y Cambios Climaticos Durante el Cuaternario*, eds. T. A. Campos and A. Perez-Gonzalez, Centro de Ciencas Medioambientales, Madrid.

Ruiz Zapata, B., Gil Garcia, M. J. and Dorado Valino, M. (1996). Climatic changes in the Spanish central zone during the last 3,000 B.P. based on polinic analysis. In *NATO ASI Series: Diachronic Climatic Impacts on*

Water Resources, eds. A. N. Angelakis and A. Issar, pp. 9–23. Berlin: Springer.

Ryan, W. and Pitman, W. (1998). *Noah's Flood. The New Scientific Discoveries about the Event that Changed History*. New York: Simon and Scuhuster, 319 pp.

Rychagov, G. I. (1997). Holocene oscillations of the Caspian Sea, and forecasts based on paleogeographical reconstructions. *Quaternary International*, 41/42, 167–172.

Sadler, J. P. and Grattman, J. P. (1999). Volcanoes as agents of past environmental change. *Global and Planetary Change*, 21, 181–196.

Sadori, L. and Narcisi, B. (2001). The postglacial record of environmental history from Lago di Pergusa, Sicily. *The Holocene*, 11, 655–670.

Sakaguchi, Y. (1982). Climatic variability during the Holocene Epoch in Japan and its causes. *Bulletin of the Department of Geography, University of Tokyo*, 14, 1–27.

Sakaguchi, Y. (1983). Warm and cold stages in the past 7600 years in Japan and their Global Correlation. *Bulletin of the Department of Geography, University of Tokyo*, 15, 1–31.

Sakaguchi, Y. (1987). Paleoenvironments in Palmyra district during the Late Quaternary. In *Paleolithic Site of the Douara Cave and Paleogeography of Palmyra Basin in Syria*, Part IV: *1984 Excavations*, eds. T. Akazawa and Y. Sakaguchi, pp. 1–27. Tokyo: University of Tokyo Press.

Sakaguchi, Y. (1989). Some pollen records from Hokkaido and Sakhalin. *Bulletin of the Department of Geography, University of Tokyo*, 21, 1–17.

Sallas, L. (1992). Propuesta de modelo climatico para el Holoceno en la vertiente Cantabrica en base los datos polinicos. *Cuaternario y Geomorfologia*, 6, 63–69.

Sanlaville, P. (1989). Considération sur l'évolution de la basse Mesopotamia au cours des derniers millénaires. *Paléorient*, 15, 5–27.

Sanlaville, P. (1992). Changements climatiques dans la Peninsula Arabique durant le Pleistocene Suprieur et L'Holocene. *Paléorient*, 18, 5–26.

Sarnthein, M., Tetzlaff, G., Koopman, B., Wolter, K. and Pflaumann, U. (1981). Glacial and interglacial wind regimes over the eastern subtropical Atlantic and NW Africa. *Nature*, 293, 193–196.

Sarnthein, M., Winn, K. and Zahn, R. (1987). Paleoproductivity of oceanic upwelling and the effect on atmospheric CO_2 and climatic change during deglaciation times. In *Abrupt Climatic Change*, eds. W. H. Berger and L. D. Labeyrie, pp. 311–337. Dordrecht: Reidel.

Savoskul, O. S. and Solomina, O. N. (1996). Late Holocene glacier variations in the frontal and inner ranges of the TianShan, Central Asia. *The Holocene* 6, 25–35.

Schaeffer, C. F. A. (1968). Commentaires sur les lettres et documents troives dans les bibliotheques prevees d'Ugarit. In *Ugaritica* V, pp. 607–768. Paris: Imprimerie Nationale et Librairie Orientaliste Paul Geuthner, 806 pp.

Schaub, R. T. (1982). The origins of the Early Bronze Age walled town culture of Jordan. In *Studies in the History and Archaeology of Jordan I*, ed. A. Hadidi, pp. 67–75. Amman: Department of Antiquites.

Schilman, B., Almogi-Labin, A., Bar-Matthews, M., Labeyrie, L., Paterne, M. and Luz, B. (2001). Long- and short-term carbon fluctuations in the Eastern Mediterranean during the late Holocene. *Geology* 29, 1099–1102.

Schilman, B., Bar-Matthews, M., Almogi-Labin, A. and Luz, B. (2002). Global climate instability reflected by Eastern Mediterranean marine records during the late Holocene. *Palaeogeography, Palaeoclimatology and Palaeoecology* 176, 157–176.

Schoell, M. (1978). Oxygen isotope analysis on authigenic carbonates from Lake Van sediments and their possible bearing on the climate of the past 10,000 years. In *The Geology of Lake Van*, eds. E. T. Degens and F. Kurtman, pp. 92–97. Ankara: The Mineral Research and Exploration Institute of Turkey.

Schove, D. J. (1984). Sunspot cycles and global oscillations. In *Climatic Changes on a Yearly to Millennial Basis*, eds. N. A. Mörner and W. Karlen, pp. 257–259. Dordrecht: Reidel.

Scott, L. and Thackeray, J. F. (1987). Multivariate analysis of late pleistocene and Holocene pollen spectra from Wonderkrater, Transvaal, South Africa. *South African Journal of Science*, 83, 93–98.

Scuderi, L. A. (1990). Tree-ring evidence for climatically effective volcanic eruptions. *Quaternary Research*, 34, 67–85.

Scuderi, L. A. (1993). A 2000-year tree ring record of annual temperatures in the Sierra Nevada mountains. *Science*, 39, 1433–1436.

Sernander, R. (1894). *Studier öfver den Gotländska, Vegetationens Utvckling-shistoria*. Uppsala: Akademisk Afhandl.

Serrano Candas, E. and Agudo Garrido, D. (1988). La deglaciacion del Valle de los Ibones Azules (Panticosa): estudio glaciomorfologico. *Cuaternario y Geomorfologia*, 2, 115–117.

Serre-Bachet, F. and Guiot, J. (1987). Summer temperature changes from tree rings in the Mediterranean area during the last 800 years. In *NATO ASI Series C*, vol. 216: *Abrupt Climatic Change*, eds. W. H. Berger and L. D. Labeyrie, pp. 89–97. Dordrecht: Reidel.

Servant, M. (1973). Séquences Continentales et Variations Climatiques: Évolution du Bassin du Tchad an Cénozoique Superieur. DSc. Thesis, University of Paris.

Servant, M. (1974). Les variations climatiques des régions intertropicals du continent Africain depuis la fin du Pleistocène. In *13th Symposium on Hydrolics*, ch. 8, pp. 1–10. Paris: Societe Hydrotechnique de France.

Servant, M. and Servant-Vildary, S. (1980). L'environnement Quaternaire du bassin du Tchad. In *The Sahara and the Nile*, eds. M. A. J. Williams and H. Faure, pp. 133–162. Rotterdam: Balkema.

Servant, M., Servant, S., Camouze, J. P., Fontes, J. C. and Maley, J. (1976). Paleolimnologie des lacs du Quaterenaire recent du bassin du Tchad. In *2nd International Symposium on Paleolimnology: Interpretation Paleoclimatique*, Mikolayki, Poland, p. 23.

Shackelton, N. J. and Opdyke, N. (1973). Oxygen isotope paleomagnetic stratigraphy of equatorial pacific core V28–238: oxygen isotope temperature and ice volumes on a 105 and 106 year scale. *Quaternary Research*, 3, 39–55.

Sharon, M. (1976). The processes of destruction and nomadism in the Land of Israel under the Moslem regime (663–1517). In *Papers in the History of Israel under the Moslem Regime*, ed. M. Sharon, pp. 7–34. Jerusalem: Yad Ben-Zvi Publishing (in Hebrew).

Shi Xingbang (1991). Natural environment of the Neolithic Age in China. In *Quaternary Geology and Environment in China, Series of the XIII INQUA Congress*, ed. Liu Tungsheng, pp. 150. Beijing: Congress Science Press.

Sivan, D. (1982). Paleogeography of the Akko Area in Holocene period. In *Third Annual Coastal Conference*, Haifa, pp. 51–63.

Sneh, A., Weissbrod, T., Ehrlich, S., Horowitz, A., Moshkovitz, S. and Rosenfeld, A. (1986). Holocene evolution of the northern corner of the Nile delta. *Quaternary Research*, 26, 194–206.

Sombroek, W. G. and Zonneveld, I. S. (1971). *Ancient Dune Fields and Fluviatile Deposits in the Rima-Sokoto Basin (N. W. Nigeria)*. [Soil Survey Paper 5] Wageningen, the Netherlands: The Soil Survey Institute.

Sonntag, C., Thorwheihe, U., Rudolph, J. *et al.* (1980). Isotopic identification of Saharian groundwaters, groundwater formation in the past. *Palaeoecology of Africa*, 12, 159–171.

Soriano, A. and Calvo, M. J. (1987). Caracterisicas, datacion y evolucion de los valles de fondo de las inmediaciones de Zaragoza. *Cuaternario y Geomorfologia*, 1, 283–293.

Stager, L. E. (1971). Appendix: Climatic conditions and grain storage in the Persian period. *The Biblical Archaeologist*, xxxiv(3), 87–88.

Stanley, D. J. and Warne, A. G. (1993). Nile delta: recent geological evolution and human impact. *Science*, 260, 628–634.

Stanley, D. J., Goddio, F. and Schnepp, G. (2001). Nile flooding sank two ancient cities. *Nature*, 412, 293–294.

Starkel, L. (1984). The reflection of abrupt climatic changes in the relief and sequence of continental deposits. In *Climatic Changes on a Yearly to Millennial Basis*, eds. N. A. Mörner and W. Karlen, pp. 135–146. Dordrecht: Reidel.

Starkel, L. B. (1991). The Vistula River valley: a case study for Central Europe. In *Temperate Paleohydrology: Fluvial Processes in the Temperate Zone During the last 15000 Years*, eds. L. Starkel, K. J. Gregory and J. B. Thornes, pp. 171–188. Chichester, UK: John Wiley & Sons.

Starkel, L. B. (1998). Frequency of extreme hydroclimatically-induced events as a key to understanding environmental changes in the Holocene. In *Water, Environment and Society in Times of Climate Change*, eds. A. Issar and N. Brown, pp. 43–67. Dordrecht: Kluwer.

Steenbeek, R. (1990). On the balance between wet and dry. Vegetation horizon development and prehistoric occupation; a paleoaecological-micromorphological study in the Dutch river area. Thesis. Free University, Amsterdam.

Stern, M. (1986). The land of Israel in the Hellenistic period and the Israelite settlement period. In *The History of the Land of Israel*, vol. 1, ed. Y. Ripel, pp. 254–223. Tel Aviv: Israeli Ministry Publishing (in Hebrew).

Stevens, L. R., Wright, H. E. Jr, and Ito, E. (2001). Proposed changes in seasonality of climate during the Lateglacial and Holocene at Lake Zeribar. *The Holocene*, 11, 747–755.

Stevenson, A. C., Macklin, M. G., Benavete, J. A. *et al.* (1991). Cambios ambientales durante el Holoceano en el valle medio del Ebro: sus implicaciones arqueologicas. *Cuaternario y Geomorfologia*, 5, 149–169.

Stiller, M. and Hutchinson, G. E. (1980). The waters of Meron: a study of Lake Huleh. VI Stable isotopic composition of carbonates of a 54 m core; paleoclimatic and paleotrophic implications. *Archiv für Hydrobiologie*, 89, 275–302.

Stiller, M., Ehrlich, A., Pollinger, U., Baruch, U. and Kaufman, A. (1983–84). *The Late Holocene Sediments of Lake Kinneret (Israel): Multidisciplinary Study of a 5 m core*. Jerusalem: Geological Survey of Israel, Ministry of Energy and Infrastructure, Jerusalem, pp. 83–88.

Stine, S. (1994). Extreme and persistent drought in California and Patagonia during medieval time. *Nature*, 369, 546–549.

Stine, S. (1998). Medieval climatic anomaly in the Americas. In *Water, Environment and Society in Times of Climate Change*, eds. A. Issar, and N. Brown, pp. 43–67. Dordrecht: Kluwer.

Street, F. A. (1979). Late Quaternary precipitation estimates for the Ziway, Shala Basin, Southern Ethiopia. *Palaeoecology of Africa*, 10/11, 135–143.

Street, F. A. and Grove, A. T. (1976). Environmental and climatic implications of late Quaternary lake-level fluctuations in Africa. *Nature*, 261, 385–390.

Street-Perrott, F. A. and Perrott, A. R. (1990). Abrupt climate fluctuations in the tropics: the influence of Atlantic Ocean circulation. *Nature*, 343, 607–612.

Street-Perrott, F. A. and Roberts, N. (1983). Fluctuations in closed basin lakes as an indicator of past atmospheric circulation patterns. In *Variations in the Global Water Budget*, eds. F. A. Street-Perrott, M. Beran and R. Ratcliffe, pp. 331–345. Dordrecht: Reidel.

Street-Perrott, F. A., Roberts, N. and Metcalfe, S. (1985). Geomorphic implications of late Quaternary hydrological and climatic changes in the Northern Hemisphere tropics. In *Environmental Change and Tropical Geomorphology*, eds. I. Douglas and T. Spencer, pp. 165–183. London: Allen and Unwin.

Striem, H. L. (1985). Quantitative and qualitative aspects of the recent climatic fluctuations. *Israel Journal of Earth Sciences*, 34, 47–48.

Sultan Hamid and Gaofa Gong (1990). Investigations on the influence of climate change on peasant rebellions in 17th century China. In *Climatic Changes and Their Impacts, Beijing International Symposium on Climatic Change*, 9–12 August, p. A-17. Beijing: Natural Science Foundation of China (CNSF).

Sun Xingjun and Chen Yinshuo (1991). Holocene palynological research in China. In *Quaternary Geology and Environment in China, Series of the XIII INQUA Congress*, ed. Liu Tungsheng, pp. 214–227. Beijing: Congress Science Press.

Suzuki, Y. (1979). 3500 years ago: climatic change and ancient civilizations. *Bulletin of the University of Tokyo Department of Geography* Univ. Tokyo, 11, 43–58.

Svensson, G. (1988). Bog development and environmental conditions as shown by the stratigraphy of Store Mosse mire in southern Sweden. *Boreas*, 17, 89–11.

Swezey, C., Lancaster, N., Kocurek, G. *et al.* (1999). Response of aeolian systems to holocene climatic and hydrologic changes on the northern margin of the Sahara: a high resolution record from the Chott Raharsa basin, Tunisia. *The Holocene*, 9, 141–147.

Szeicz, J. M. and Macdonald, G. M. (1996). A 930-year ring-width chronology from moisture-sensitive white spruce (*Picea glauca* Moench) in northwestern Canada. *The Holocene*, 6, 345–351.

Takahashi, K. (1987). An analysis of long term variation of storm damage in Japan. In *The Climate of China and Global Climate, Beijing International Symposium on Climate*, 30 October to 3 November 1984, eds. Ye Dhuzheng, Fu Congbin, Chao Jiping and M. Moshino, pp. 13–19. Beijing: China Ocean Press and Springer Verlag.

Talbot, M. R. (1980). Environmental responses to climatic change in the West African Sahel over the past 20,000 years. In *The Sahara and the Nile*, eds. M. A. J. Williams and H. Faure, pp. 37–62. Rotterdam: Balkema.

Talma, A. S. and Vogel, J. C. (1992). Late Quaternary paleotemperatures derived from a speleothem from Cango Caves, Cape Province, South Africa. *Quaternary Research*, 37, 203–213.

Tang Lingyu, Shen Caiming and Yu Ge (1990). Preliminary study on the 7500–5000 B.P. climatic sequence in the middle and lower beaches of the Yangtze River. In *Climatic Changes and Their Impacts, Beijing*

International Symposium on Climatic Change, 9–12 August, pp. A21. Beijing: Natural Science Foundation of China (CNSF).

Taylor, K. C., Lamorey, G. W., Doyle, R. B. *et al.* (1993). The 'flickering switch' of late Pleistocene climate change. *Nature*, 361, 432–436.

Thomas, M. F. and Thorp, M. B. (1995). Geomorphic response to rapid climatic and hydrologic change during the Late Pleistocene and early Holocene in the humid and sub-humid tropics. *Quaternary Science Review*, 14, 193–207.

Thompson, F. H. (ed.) (1980). *Archaeology and Coastal Change*. London: The Society of Antiquitaries of London, 154 pp.

Thompson, L. G., Mosley-Thompson, E., Davis, M. E. *et al.* (1989). Holocene–Late Pleistocene climatic ice core records from Qinghai-Tibetan plateau. *Science*, 246, 474–477.

Thunell, R. C. and Williams, D. F. (1983). Paleotemperature and paleosalinity history of the Eastern Mediterranean during the Late Quaternary. *Palaeogeography, Palaeoclimatology and Palaeoecology*, 44, 23–39.

Thunell, R. C. and Williams, D. F. (1989). Glacial-Holocene salinity changes in the Mediterranean Sea: hydrographic and depositional effects. *Nature*, 338, 493–496.

Trump, D. H. (1990). *Malta: An Archaeological Guide*. Valetta, Malta: Progress Press, 172 pp.

Tsoar, H. (1995). Desertification in northern Sinai in the eighteenth century *Climatic Change*, 29, 429–438.

Tsoar, H. and Goodfriend, G. A. (1994). Chronology and paleoenvironmental interpretation of Holocene aeolian sands at the inland edge of the Sinai–Negev erg. *The Holocene* 4, 244–250.

Tsukada, M. (1986). Vegetation in prehistoric Japan, the last 20,000 years. In *Windows on the Japanese Past: Studies in Archaeology and Prehistory*, ed. R. J. Pearson, pp. 11–56. Ann Arbor: University of Michigan Press.

Tyson, P. D. (1986). *Climatic Change and Variability in Southern Africa*. Cape Town: Oxford University Press.

Tyson, P. D. and Lindesay, J. A. (1992). The climate of the last 2000 years in southern Africa. *The Holocene*, 2, 271–278.

Tzafrir, Y. (1984). The Arabic conquest and the process of settlement deterioration of the Land of Israel. *Katedra*, 32, 69–74 (in Hebrew).

Ussishkin, D. (1986). The land of Israel during the Chalcolithic period. In *The History of the Land of Israel*, vol. 1, ed. Y. Ripel, pp. 47–60. Tel Aviv: Israel Defense Ministry Publishing (in Hebrew).

Van Campo, E., Duplessy, J. C. and Rossignol-Strick, M. (1982). Climatic conditions deduced from a 150-kyr oxygen isotope-pollen record from the Arabian Sea. *Nature*, 296, 56–59.

Van den Brink, L. M. and Janssen, C. R. (1985). The effect of human activities during cultural phases on the development on montane vegetation in the Serra da Estrela, Portugal. *Review of Paleobotany and Palynology*, 44, 193–215.

Van de Plassche, O. (1982). Sea level changes and water level movements in the Netherlands during the Holocene. Thesis, Free University of Amsterdam.

Van der Valk, L. (1992). Coastal evolution in the beach-barrier area of the Western Netherlands. PhD Thesis, Free University of Amsterdam.

van der Knaap, W. O. and van Leeuwen, J. F. N. (1995). Holocene vegetation succession and degradation as responses to climatic change and human activity in the Serra de Estrela, Portugal. *Review of Paleobotany and Palynology*, 64, 153–211.

Van der Woude, J. D. (1983). Holocene paleoenvironmental evolution of a perimarine fluviatile area. PhD Thesis, Free University of Amsterdam and Analecta Praehistorica Leidensia XVI, Leiden University Press, 101 pp.

Van Geel, B. and Renssen, H. (1998). Abrupt climate change around 2,650 B.P. in north west Europe: evidence for climatic telecommunications and tentative explanation. In *Water, Environment and Society in Times of Climate Change*, eds. A. S. Issar and N. Brown, pp. 21–42. Dordrecht: Kluwer.

Van Zeist, W. (1969). Reflections on prehistoric environments in the Near East. In *The Domestication and Exploitation of Plants and Animals*, eds. P. J. Ucko, and G. W. Dimbleby, pp. 35–46. London: Duckworth.

Van Zeist, W. (1980). Vegetational and climatic interpretation of palynological data. In *Proceedings of the Bat Sheva Seminar on Approaches and Methods in Paleoclimatic Research, with Emphasis on Rich Areas*, Jerusalem and Rehovot, 20–30 October, pp. 69–71.

Van Zeist, W. and Bakker-Heeres, J. A. H. (1979). Some economic and ecological aspects of the plant husbandry of Tell Aswad. *Paléorient*, 5, 161–169.

Van Zeist, W. and Bottema, S. (1977). Palynological investigations in Western Iran. *Paleo-Historia*, 19, 19–85.

Van Zeist, W. and Bottema, S. (1982). Vegetational history of the Eastern Mediterranean and the Near East during the last 20,000 years. In *British Archaeological Reports* International Series 133: *Palaeoclimates, Palaeoenvironments and Human Communities in Eastern Mediterranean Region in Later Prehistory*, pp. 277–321. Oxford: British Archaeological Reports.

Van Zinderen Bakker, E. M. (1976). The evolution of Late Quaternary paleoclimates of Southern Africa. *Paleoecology of Africa*, 9, 160–202.

Versoub, K. L., Fine, P., Singer, M. J. and TenPas, L. (1993). Pedogenesis and paleoclimate: interpretation of the magnetic susceptibility record of Chinese loess–paleosol sequences. *Geology*, 21, 1011–1014.

Wang Jian, Lu Houyuan and Shen Caiming (1990). 'Factor Interpretation Method' and quantitative analysis of climatic changes over the last 12 000 years. In *Regional Conference on Asian Pacific Countries of I.G.U., China Global Change and Environmental Evolution in China*, Section III, eds. Liu Chuang, Zhao Songqiao, Zhang Peiyuan and Shi Peijun, pp. 62–68. Hohot. P. R. China: Editorial Board of Arid Land Resources.

Wan Jingtai, Derbyshire, E., Meng Xingnim and Ma Jinhui (1991). Natural hazards and geological processes: an introduction to the history of natural hazards in the Gansu Province, China. In *Quaternary Geology and Environment in China, Series of the XIII INQUA Congress*, ed. Liu Tungsheng, pp. 285–296. Beijing: Congress Science Press.

Wang Shaowu (1990). The climatic charateristics of the Little Ice Age in China, Beijing In *Climatic Changes and Their Impacts, Beijing International Symposium on Climatic Change*, 9–12 August, p. A-7. Beijing: Natural Science Foundation of China (CNSF).

Wang Shaowu *et al.* (1987). Drought/flood variations for the last 2000 years in China in comparison with the global climatic change. In *The Climate of China and Global Climate, Beijing International Symposium on Climate*, 30 October to 3 November 1984, eds. Ye Dhuzheng, Fu Congbin, Chao Jiping and M. Moshino, pp. 20–29. Beijing: China Ocean Press and Springer Verlag.

Wang Sumin and Li Jianren (1991). Late Cenozoic lake sediments in China. In *Quaternary Geology and Environment in China, Series of the XIII INQUA Congress*, ed. Liu Tungsheng, pp. 49–57. Beijing: Congress Science Press.

Weinstein, M. (1976). The late Quaternary vegetation of the northern Golan. *Pollen and Spores*, 18, 553–562.

Weinstein, J. (1981). The Egyptian Empire in Palestine: a reassessment. *Bulletin of the American Schools of Oriental Research*, 241, 1–28.

Weisrock, A. (1980). The littoral deposits of the Saharian Atlantic Coast since 150,000 years B. P. *Palaeoecology of Africa*, 12, 277–287.

Weiss, B. (1982). The decline of Late Bronze age civilization as a possible response to climatic change. *Climatic Change*, 4, 173–198.

Weiss, H., Coutry, M. A., Wetterstrom, W. *et al.* (1993). The genesis and collapse of third millennium north Mesopotamian civilization. *Science*, 261, 995–1004.

Wen Qizhong and Qiao Yulo (1990). Preliminary probe of climatic sequence in the last 13000 years in Xinjiang region. In *Climatic Changes and Their Impacts, Beijing International Symposium on Climatic Change*, 9–12 August, p. c-18. Beijing: Natural Science Foundation of China (CNSF).

Wenlin, Z. (1988). *The History of Chinese population*. Beijing: People's Publishing House.

Whitmore, J. T., Brenner, M., Curtis, J. H., Dahlin, B. H. and Leyden, B. W. (1996). *The Holocene*, 6, 273–287.

Wreschner, E. E. (1977). Sea level changes and settlement location in the coastal plain of Israel during the Holocene. In *Moshe Stekelis Memorial Volume*, eds. B. Arensburg, and O. Bar-Yosef, pp. 260–271. Jerusalem: The Israel Exploration Society.

Wright, G. E. (1938). The chronology of Palestine pottery in middle Bronze I. *Bulletin of American Schools of Oriental Research*, 71, 27–34.

Wright, H. E. Jr (1966). Stratigraphy of lake sediments and the precision of the palaeoclimatic record. In *Symposium of the Royal Meteorological Society*, London, pp. 158–173.

Wright, H. E. Jr (1976). The environmental setting for plant domestication in the Near East. *Science*, 194, 385–389.

Wu Xiangding and Zhan Xuzhi (1991). On extracting proxy data of climatic change from tree-ring width in China. In *Quaternary Geology and Environment in China, Series of the XIII INQUA Congress*, ed. Liu Tungsheng, pp. 245–252. Beijing: Congress Science Press.

Xiong Dexin (1990). An approach to autumn rain in the Hunan Province during the last 1300 years. In *Climatic Changes and Their Impacts*,

Beijing International Symposium on Climatic Change, 9–12 August, p. A-4. Beijing: Natural Science Foundation of China (CNSF).

Xue Chunting, Ye Zhizhen and He Qixiang (1991). Holocene coastal sedimentation of China. In *Quaternary Geology and Environment in China, Series of the XIII INQUA Congress*, ed. Liu Tungsheng, pp. 64–72. Beijing: Congress Science Press.

Xu Guochang and Yao Hui (1990). The changes of historical climate in the Holocene of West China. In *Climatic Changes and Their Impacts, Beijing International Symposium on Climatic Change*, 9–12 August, p. c-23. Beijing: Natural Science Foundation of China (CNSF).

Xu Xiangding and Yin Xungdang (1990). A study on the abrupt dryness–wetness changes in the middle Yellow River Valley during historical times. In *Climatic Changes and Their Impacts, Beijing International Symposium on Climatic Change*, 9–12 August, p. A-5. Beijing: Natural Science Foundation of China (CNSF).

Yakir, D., Issar, A., Gat, J., Adar, E., Trimborn, P. and Lipp, J. (1994). [13]C and [18]O of wood from the Roman siege rampart in Masada, Israel (AD 70–73): evidence for a less arid climate for the region. *Geochimica et Cosmochimica Acta*, 58, 3535–3539.

Yang Huai-Jan and Xie Zhiren (1984). Sea level changes in East China over the past 20,000 years. In *The Evolution of the East Asian Environment*, vol. I ed. R. O. Whyte., pp. 288–308. Hong Kong Center of Asian Studies, University of Hong kong.

Yang Huaijen (1991). Paleomonsoon and the mid-Holocene climatic and sea level fluctuations in China. In *Quaternary Geology and Environment in China, Series of the XIII INQUA Congress*, ed. Liu Tungsheng, pp. 326–336. Beijing: Congress Science Press.

Yang Huaijen and Chen Xiqing (1987). Quaternary transgressions, eustatic changes and movement of shorelines in North and East China. In *Proceedings of the First International Conference on Geomorphology*, Manchester, 15–21 September, pp. 807–827. Chichester, UK: John Wiley & Sons.

Yao Tandong, Yafeng Shi and Thompson, L. G. (1997). High resolution record of paleoclimate since the Little Ice Age from the Tibetan ice cores. *Quaternary International*, 37, 19–23

Yll, E. I., Perez-Obiol, R., Pantaleon-Cano, J. and Roure, J. R. (1995). Dinamica del paisaje vegetal en la vertiente Mediterranea de la Peninsula Iberica e islas Baleares el Tardiglaciar hasta el presente. In *IX Reunion Nacional Sobre Cuaternario, Reconstruccion de Paleoambientes y Cambios Climaticos durante el Cuaternario*, eds. T. A. Campos and A. Perez-Gonzalez, pp. 319–328. Madrid: Centro de Ciencas Medioambientales.

Yonekura, Y. and Ota, Y. (1986). Sea level and tectonics in the Late Quaternary. In *Recent Progress of Natural Sciences in Japan*, 11, 11–34.

Yoshino, M. M. (1978). *Climatic Change and Food Production*, pp. 331–342. Tokyo: University of Tokyo Press.

Yoshino, M. M. and Urushibara, K. (1979). Paleoclimate in Japan since the Last Ice Age. *Climatological Notes*, 22, 1–24.

Zagwijn, W. H. (1974). The paleogeograhic evolution of the Netherlands during the Quaternary. *Geologie en Mijnbouw*, 53, 369–385.

Zagwijn, W. H. and Van Staalduinen, C. J. (eds) (1975). *Toelichting bij de Geologische Overzichtskaarten van Nederland*. Haarlem: Rijks Geologie Dienst, 134 pp.

Zangvil, A. (1979). Temporal fluctuations of seasonal precipitation in Jerusalem. *Tellus*, 31, 413–420.

Zao Xitao, Lu Gyangi, Wang Shohong, Wu Xuezhong and Zhang Jingwen (1990). Holocene stratigraphy, in Qingfeng, Jinahu County, Jingsu Province, China and its reflection on the climatic and environmental changes. In *Climatic Changes and Their Impacts, Beijing International Symposium on Climatic Change*, 9–12 August, pp. 1–18. Beijing: Natural Science Foundation of China (CNSF).

Zazo, C., Goy, J. L., Lario, J. and Silva, P. G. (1996). *Littoral Zone and Rapid Climatic Changes during the last 20,000 Years*. The Iberia Study Case. Z. Geomorph. N. E. Suppl. Bd. 102, 119–134.

Zhang De'er. (1991). Climate changes in recent 1000 years in China. In *Quaternary Geology and Environment in China, Series of the XIII INQUA Congress*, ed. Liu Tungsheng, pp. 208–213. Beijing: Congress Science Press.

Zhang Lansheng, Fang Xiuqui and Shi Peijun (1990). Climatic changes and their impacts. In *Beijing International Symposium on Climatic Change*, 9–12 August. Beijing: Natural Science Foundation of China (CNSF).

Zhang Peiyuan and Ge Quansheng (1990). Abrupt climatic change: introduction and a case study. In *Climatic Changes and Their Impacts, Beijing International Symposium on Climatic Change*, 9–12 August, p. C-1. Beijing: Natural Science Foundation of China (CNSF).

Zhang Peiyuan and Wu Xiangding (1990). Regional response to global warming: a case study in China. In *World Laboratory and CCAST Workshop Series*, vol. 5, eds. Yino Yuan Li *et al.*, pp. 26–41. London: Gordon and Breach.

Zheng Benxing (1991). Studies of the Quaternary geology in Kunlun Mountains. In *Quaternary Geology and Environment in China, Series of the XIII INQUA Congress*, ed. Liu Tungsheng, pp. 396–402. Beijing: Congress Science Press.

Zhong-wei Yan, Jin-Ju Ji and Dhu Zheng-Ye (1990). Northern hemispheric summer climatic jump in the 1960s. In *Climatic Changes and Their Impacts, Beijing International Symposium on Climatic Change*, 9–12 August, p. A-20. Beijing: Natural Science Foundation of China (CNSF).

Zhou Weijian and An Zhisheng (1991). [14]C chronology of the loess plateau in China. In *Quaternary Geology and Environment in China, Series of the XIII INQUA Congress*, ed. Liu Tungsheng, pp. 192–200. Beijing: Congress Science Press.

Zhou Yuowu, Qiu Guoqing and Guo Dongxin (1991). Changes of permafrost in China during the Quaternary. In *Quaternary Geology and Environment in China, Series of the XIII INQUA Congress*, ed. Liu Tungsheng, pp. 86–94. Beijing: Congress Science Press.

Zoller, M. (1977). Alter und Ausmass postglazialer Klimatschwankungen in den Schweizer Alpen. In *Dendrochronologie und Klimatschwankungen in Europa*, ed. Frenzel, B., pp. 271–281. Wiesbaden: Steiner Verlag.

Index